NCA Report Series, Volume 6:

Scenarios for Research and Assessment of Our Climate Future:

Issues and Methodological Perspectives for the

U.S. National Climate Assessment

NCA Report Series

The National Climate Assessment (NCA) Report Series summarizes regional, sectoral, and process-related workshops and discussions being held as part of the Third NCA process.

The workshop on scenarios for research and assessment of our climate future was held in Arlington, Virginia on December 6-8, 2010. Volume 6 of the NCA Report Series summarizes the discussions and outcomes of this workshop. A list of completed and planned reports in the NCA Report Series can be found online at http://assessment.globalchange.gov.

CONTENTS

CONTENTS

PREFACE

This document reports on a workshop held in December 2010 in Arlington, Virginia to explore needs, options, and research for the development of scenarios to support science and assessment of climate and global change over the coming decades. The overall effort was led by the science community and coordinated through a research community steering group. Responding to the interagency U.S. Global Change Research Program (USGCRP), the workshop was sponsored by the U.S. Department of Energy's Office of Biological and Environmental Research. The purpose, scope, and objectives of the effort were informed by inputs from a federal coordinating committee comprised of representatives of participating USGCRP agencies. The workshop brought together leading researchers, scenario developers, stakeholders, and federal officials to examine the potential uses of scenarios in research on and assessments of climate change and response options. The results also have significant implications for climate research beyond assessments, improving understanding of the current scientific basis for scenario development and identifying methods for improving consistency in their use and interpretation. The workshop included plenary-session presentations and panels as well as breakout group discussions.

This report summarizes the insights developed through the workshop process. Chapter 1 is in the form of a letter report from the research community steering group that organized the workshop. Chapter 2 is a revised version of a white paper that was prepared to provide background context for workshop participants. Chapter 3 includes an overview of each of the presentations made during the workshop as well as summaries of the points made during breakout group discussions.

The workshop report defines key terms and establishes a conceptual framework for developing consistent scenarios across different end uses and spatial scales. It reviews the scientific underpinnings of scenarios, discusses the application of scenarios in past National Climate Assessments, and identifies potential users of and future needs for scenarios. Although not limited to this application, an immediate priority is meeting scenario needs for the next U.S. National Climate Assessment (NCA). The NCA is driven by a legislative mandate that requires preparation of a periodic report; the next of which is to be completed in 2013 (hereafter referred to as "NCA 2013"). Beyond this near-term priority, the ongoing NCA process will also require improved scenarios and corresponding methodologies, supporting NCA participants that include federal agencies; regional, state, and local governing bodies; academia; business associations; and nongovernmental organizations. This document explores resources that could be developed and made freely available to those groups, and it considers potential approaches for providing methods, data, and other tools for Assessment participants. The report also surveys recent scientific advances in the development and use of scenarios and current activities in the research community that could provide needed inputs.

The report does not reflect a consensus among participants but rather a description of needs, options, and challenges as seen by the participants in their individual capacities.

Chapter 1:
Letter Report on Needs and Options for Scenarios for the National Climate Assessment

Richard Moss, Joint Global Change Research Institute, Pacific Northwest National Laboratory; Linda Mearns, National Center for Atmospheric Research; Molly Cross, Wildlife Conservation Society; Holly C. Hartmann, University of Arizona; Kathy Hibbard, Pacific Northwest National Laboratory; Robert Lempert, RAND Corporation; Philip Mote, Oregon State University; Edward Parson, University of Michigan; Rich Richels, Electric Power Research Institute; John Robinson, University of British Columbia; Cynthia Rosenzweig, NASA Goddard Institute for Space Studies; Joel Smith, Stratus Consulting; and Tom Wilbanks, Oak Ridge National Laboratory

1.1 Introduction

This chapter of the workshop report was prepared at the request of the Integrated Assessment Research Program of the U.S. Department of Energy, Office of Biological and Environmental Research, by an ad hoc group of researchers who helped organize the workshop. The workshop, conducted in support of the U.S. Global Change Research Program (USGCRP), was held on December 6–8, 2010 in Arlington, Virginia. The primary focus of this meeting was to review current practices and identify options for use of scenarios in the next U.S. National Climate Assessment (NCA). Although the immediate intended audience of this report is the National Climate Assessment Development and Advisory Committee (NCADAC), the federal advisory committee leading the development of the Assessment, the workshop was planned and the report written to support a broader range of research community interests in climate change science, analysis, and planning.

This chapter of the workshop report summarizes information from a background white paper (Chapter 2) and the workshop discussions (Chapter 3) to identify options for supporting NCA objectives. These objectives address the immediate task of the NCADAC to prepare a report scheduled for completion in 2013, as well as the long-term goal of developing resources to support future Assessments. This chapter describes different types of scenarios in global change research and briefly reviews their use in past U.S. National Climate Assessments. It provides insight into options for types of scenarios that could be used and highlights recent developments in scenarios research and practices that create new opportunities for the NCADAC to consider. The insights reflect the emergent themes and diverse perspectives of the members of the ad hoc steering group and the approximately 70 workshop participants who provided the foundation for the report.

1.2 Scenarios and the National Climate Assessment

The U.S. Global Change Research Program is conducting a climate assessment for the United States. The NCA will (1) produce a report expected to be completed in 2013, and (2) develop products and processes to support ongoing distributed assessment activities in regions and sectors across

the country. This two-pronged assessment approach defines needs for a scenario strategy that supports both coordinated synthesis for the 2013 report and adaptable scenario products and processes to inform a sustained Assessment process.

The 2013 report will apply current scientific understanding to regional, sectoral, and crosscutting issues in a comprehensive assessment of potential impacts, adaptation, and vulnerability to climate change within a context of how communities and the Nation as a whole can work to create sustainable and environmentally sound development pathways. The report will need to synthesize the results of assessment activities undertaken by participants in regions and sectors across the United States. Quantitative and qualitative scenarios of emissions, climate, and related environmental conditions can provide common assumptions and a means for coordinating scientific information provided to assessment teams, thus aiding synthesis. However, in addition to establishing common or shared assumptions, the scenarios need to be developed in such a way that helps assessment teams address regional, sectoral, and crosscutting issues of greatest concern to their stakeholders.

These immediate purposes and needs of NCA 2013 are embedded in a broader, long-term strategy in which assessments also serve as a key mechanism for applying science to place- or sector-specific questions in order to foster dialogue and learning between the research community and stakeholders. Thus, in addition to preparing the 2013 report, the NCA process is also seeking to develop resources to support ongoing distributed assessments that will draw upon the work of scientists and stakeholders across the country. This second, broader purpose has relevance for the types of scenario resources that need to be developed. In this context, the goal of working with scenarios is not to predict the future, but to better understand the implications of uncertainties for decisions involving valued attributes or activities that could be affected by climate change. Scenarios of future climate and related conditions are used by assessors and stakeholders in order to consider how robust different development, adaptation, or mitigation options may be given current scientific understanding of the range of potential climate change and relevant ecological and socioeconomic factors. To facilitate this application of scenarios, it is necessary to create and make available tools and resources that encourage regional or sectoral groups to develop their own scenarios nested within a broad range of climate and socioeconomic futures. Among the resources that support and facilitate stakeholder and public engagement are decision theaters, visualization techniques, and strategies for employing scientific information in participatory dialogues.

1.3 Types of Scenarios used in Global Change Research and Assessment

The term "scenarios" describes qualitative and quantitative information about different aspects of the future developed to investigate the potential consequences of climate change. The preparatory white paper and discussions at the workshop classified scenarios according to their uses, content, and the types of models or methods used to provide inputs to them. We note that it is important to distinguish between the needs of decision makers and the ways in which scenarios are used to evaluate robust decision making from the ways that scenarios can be employed by researchers to coordinate studies across scales or sectors.

The major types of scenarios deemed relevant to the NCA include

*   *Emissions scenarios* are descriptions of potential future emissions to the atmosphere of greenhouse gases and other radiatively important gases and particles that are used to explore the implications of alternative energy and technology futures. Emissions scenarios are also used as inputs to climate models. They are not forecasts or predictions. They focus on long-term (e.g., decades to centuries) trends in energy and land-use patterns, not short-term fluctuations. Emissions scenarios are often used as inputs to develop scenarios of future radiative forcing for climate models.

*   *Climate scenarios* are plausible representations of future climate conditions (temperature, precipitation, and other variables) produced using a variety of techniques including scaling of observed climate, spatial and temporal analogs (in which climates from other locations or periods are used as illustrative future conditions), and mathematical models. Regional-scale climate scenarios and projection methods for impact and adaptation assessment are highly relevant for the NCA. Climate scenarios are often used as inputs to models of the impacts of climate change.

*   *Environmental scenarios* focus on changes in environmental conditions such as water availability and quality, sea-level rise (incorporating geological and climate drivers), land cover, land use, and air quality. These factors can vary as a result of climate change, or as a result of other driving forces such as human settlement or resource extraction. The potential impact of climate change and the effectiveness of adaptation options cannot be understood without examining interactions with broader environmental conditions and aspects of global change.

*   *Socioeconomic scenarios* for assessment of impacts, adaptation, and vulnerability project future demographic, economic, institutional, and other characteristics needed for different types of impact modeling and research. This information is important for projecting emissions and assessing the sensitivity and adaptive capacity of society and how different patterns of economic growth and social change could affect sensitivity and adaptation in the future.

*   *Narratives* describe in qualitative form political, institutional, and other factors that influence future forcing, vulnerability, and responses. They can describe the overall logic embedded in a quantitative scenario of socioeconomic factors or emissions and are based on analysis of and extrapolation from current conditions and historical experience. Narratives are useful because some socioeconomic factors affecting emissions and vulnerability are not effectively quantified (e.g., institutions). Narratives can facilitate coordination across spatial scales and substantive domains. The non-climate scenario literature often uses the term "narratives" to refer to qualitative descriptions or stories about the future that are strategically developed to engage the imaginations of decision makers (end users) and lead them to consider issues that they might otherwise neglect but that are nonetheless important.

Other typologies of scenarios are also available and valuable to consider because they offer guides to the potential use of the scenarios. These include a categorization of types of scenarios by their use for informing decision making: e.g., scenarios constructed to help decision makers understand the most important drivers of their future and how best to respond ("intuitive logics"); scenarios in which desirable futures are defined and the scenarios specify how these visions might be attained ("backcasting approaches"); and scenarios that build on trends but add surprise to traditional forecasting methods ("probabilistic modified trends"). Some scenarios are created by researchers and experts to portray ranges of projections (from high to low), whereas others are developed using interactive participatory processes to explore the implications of uncertainty for decisions.

Several concerns about the use of scenarios have been raised in the literature. In a set of scenarios, specific combinations of projected conditions are selected *a priori* from among a large set of combinations of model input parameters to span a range of possible futures. In fact, the probability associated with any one future is infinitesimally small since it reflects a combination of point values of key parameters. From the perspective of uncertainty characterization or quantification, there is no basis for confidence that any particular scenario set captures the range of possible combinations of climate, environmental, and socioeconomic conditions that should be considered in impacts assessment. Another concern is that users can develop overconfidence in scenarios, coming to believe that they represent the most important or likely possibilities when in fact their likelihood is quite low.

A related debate concerns whether probabilities can be usefully associated with scenarios. The motivation for providing probabilities is that without quantification of relative likelihoods, decision makers will have insufficient information upon which to base decisions or will develop their own assessments of relative likelihood that depart from the best judgment of experts. Debate centers on whether the resulting estimates may overstate existing knowledge, under-represent uncertainty, or whether attaching probabilities to scenarios is at odds with their proper use in decision-making contexts. It is essential that scenarios prepared for use in the NCA be accompanied by clear guidance on their interpretation, uses, and limits.

These issues and concerns notwithstanding, participants in the workshop generally felt that

use of scenarios was an important option for the NCADAC to use for both the 2013 report and to make available for the long-term, sustained Assessment process. The strength of the scenario approach is that research and assessment results are more easily communicated to stakeholders in the context of scenarios, and that a scenario planning process implemented in sectoral and place-based contexts enables decision makers to analyze the implications of different levels and rates of climate change for decision options. The scenario approach also facilitates coordination across the many research communities involved in climate research and assessment, and through identification of a manageable number of discrete scenarios, is computationally and analytically efficient and avoids the combinatorial problems associated with the range of possible futures. For these reasons, a mix of scenarios and scenario-based methodologies is likely to be of use to the NCA 2013 and future products.

For the purposes of discussion of options for scenarios in the NCA, it is useful to develop a conceptual framework that helps clarify the relationships among different types of scenarios, models, and other data and information. There is a danger that without a clear set of definitions, the term scenario will become too vague and thus not helpful in differentiating among classes of products. As illustrated in Figure 1.1, when one considers relationships across research disciplines, the production of scenarios by one type of model as input assumptions into another is clear, especially in their traditional linear relationship.

A number of types of models are relevant to the NCA and are used in the development and application of scenarios. Earth system and climate models are systems of differential equations—based on the basic laws of physics, fluid motion, and chemistry—that represent Earth's climate system on a three-dimensional grid that extends through the land, ocean, and atmosphere. They use scenarios of emissions, radiative forcing, land cover, and other factors as inputs. The latest generation of high-resolution models has improved representation of fine-scale climate and Earth system processes but still has relatively coarse spatial resolutions (hundreds of kilometers). Regional climate models provide some of the needed resolution over smaller spatial domains; however, since the regional models are driven by the general climate simulated by global models, they still lack the skill to project

Scenarios, Modeling, and Use: Dynamic Interactions for the NCA

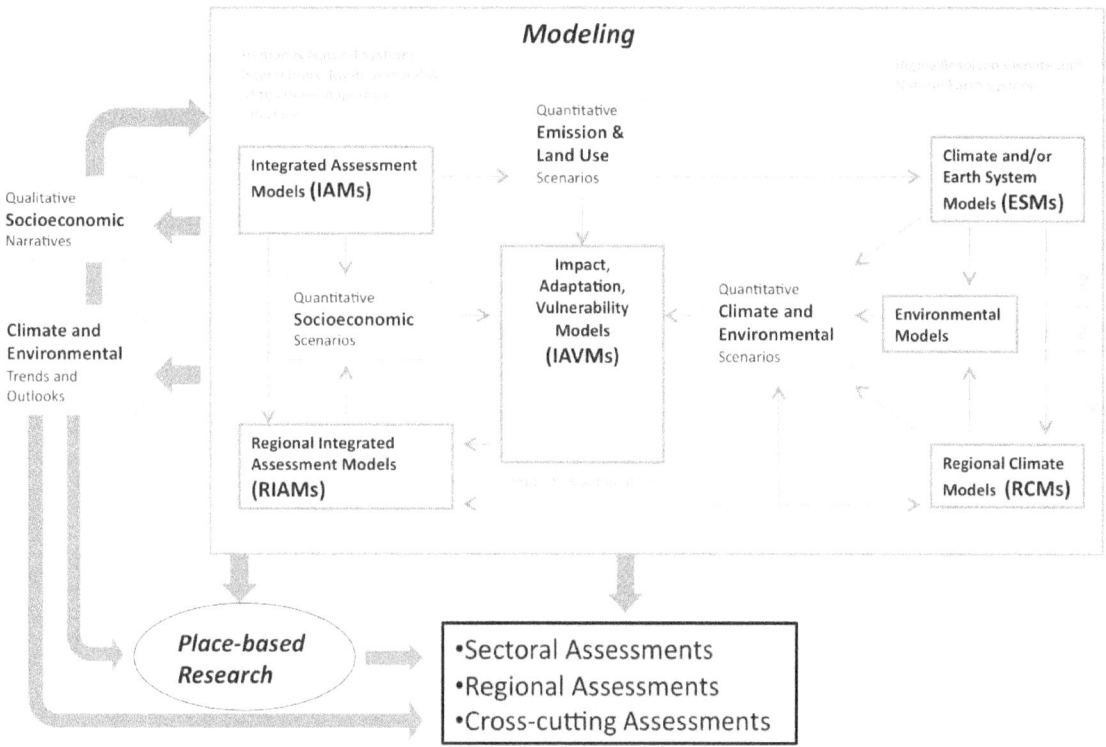

Figure 1.1. Scenarios, modeling, and use: dynamic interactions for the NCA. The figure illustrates the likely flows of information among groups of models and clarifies when models and other techniques are being used to develop scenarios that can be used as inputs to another class of models. Blue colored shapes indicate models, yellow scenarios or narratives, and green end-point assessments. See the text for additional information about the types of models depicted in the figure (Courtesy of Bob Vallario, John Hall, and Richard Moss).

change accurately. Other methods are also available for downscaling climate information, each with their own appropriate uses and limits.

Impact, adaptation, and vulnerability models include a wide range of models that have been developed to study the interactions of environmental and managed systems with climate through various techniques including process and mechanistic simulation, statistical and empirical relationships, and time series analyses. They include models of managed systems such as agriculture, air quality, human health, and markets for climate-sensitive commodities as well as models of diverse environmental systems. Environmental models include a variety of biophysical process models such as hydrology models, geomorphologic process models, vegetation and species response models, fire models, and coastal inundation models that can be used to study the functioning of these systems or processes, their interactions with climate, and in some cases to generate environmental scenarios that

are inputs into impacts, vulnerability, or adaptation models. Most impact, adaptation, or vulnerability models use climate scenarios based on global or regional climate model results and other methods including, in some cases, socioeconomic scenarios. Integrated assessment models simulate the interactions between human systems (including decision options) and the natural components of the overall Earth system. Historically, they have been used to develop greenhouse gas emissions scenarios, usually over the 21st century. In recent years, they have undergone transformation to enable insights into land use (with implications for carbon, nitrogen, and water cycles), socioeconomic scenarios, and importantly, interacting systems and stressors spanning impacts, adaptation, or vulnerability domains. For more detailed information on the above models, see the related report on models for the NCA.

In the NCA, when global model outputs are downscaled for use at regional, sub-regional, and

Global scale	Regional scale (continental/sub-continental)	Local scale

• *Earth system and climate model outputs*
• Global environmental models (major feedback loops involving carbon, water, and nitrogen)
• Socioeconomic scenarios (e.g., demographic projections)
• Emission scenarios
• *Integrated Assessment Models (IAM) and derived scenarios (e.g., emissions, land use)*
	• *Earth system and climate models, and derivative products*
	  o *General Circulation Models (GCMs)*
	  o *Regional Climate Models (dynamical)*
	  o <u>Statistical downscaling</u>
	  o <u>Climate outlooks</u>
	• *Environmental or biophysical process models (e.g., hydrology, marsh dynamics, fire behavior, etc.)*
	• *Impact, adaptation, and vulnerability (IAV) models; modeled impacts to natural resources or built infrastructure, human health, etc.)*
	• Environmental scenarios (e.g., sea level rise, land cover or use change)
	• Socioeconomic scenarios (including participatory processes for planning purposes)
	• *Regional IAMs*
		• Climate scenarios and <u>outlooks</u>
		• Environmental scenarios
		• *Environmental or biophysical process models*
		• Socioeconomic scenarios (including participatory processes for planning purposes)
		• *IAV models*

Scenarios appear in plain type, models in italics, and other techniques (e.g., statistical methods or expert judgment) are underlined.		

Table 1.1. Example of scenarios, models, and processing techniques available for use at different spatial scales in the National Climate Assessment

local scales, information is developed, processed, and used in a variety of ways, not only in models, but also in scaling techniques, qualitative expert judgments, and stakeholder engagement processes (Table 1). Because of the great diversity of applications, when using the term "scenario," attaching a modifier that indicates the focus or use of the scenario in either modeling (for intermediate users) or in decision support or participatory processes (for end users) will help avoid confusion.

1.4 Lessons from Past U.S. National Assessments

Two National Climate Assessments have been prepared since 1990, when the Global Change Research Act was passed. The core reports of these assessments were published in 2001 and 2009 (NAST, 2001; Karl *et al.*, 2009). The two reports and processes were very different in character and extent. The 2001 report included a concise overview report, a longer, technical "foundation" report, and detailed reports for eight major regions, five sectors, and native peoples and homelands. The 2001 report sought to establish an ongoing assessment process. The assessment published in 2009 summarized information contained in 21 "Synthesis and Assessment Products" produced by the USGCRP when it was known as the Climate Change Science Program (CCSP) from 2002–2009. The development of a long-term sustained assessment process was left to a future assessment.

A range of scenarios was developed and provided for both of these assessments. For NCA 2000, three basic categories of scenarios were developed: climate, ecosystem/vegetation, and socioeconomic. For climate, two model simulations—one developed by the Canadian Centre for Climate Modelling and Analysis (CGCM1) and the other by the Hadley Centre for Climate Prediction and Research of the Meteorological Office of the United Kingdom (HadCM2) —were recommended, both forced with the mid-range IS92a emissions baseline scenario of the Intergovernmental Panel on Climate Change (IPCC). The full range of variables relevant for analysis of impacts was available through the modeling teams. The Vegetation/ Ecosystem Modeling and Analysis Project (VEMAP) was used to generate future ecosystem scenarios for the conterminous U.S. Outputs were based on biogeochemistry models in the near term (2025–2034) and biogeography models in the longer term (2090–2099). Socioeconomic scenarios were also explicitly developed to provide context for evaluation of impacts, vulnerabilities, and adaptations. County-level projections of a few key variables relevant to all regions and sectors to 2030, and aggregate national-scale projections to 2100 were provided. In the short term (2030), three projections (high, middle, and low) depended upon varying assumptions of fertility, mortality, migration, labor-force participation, and productivity by age group. A common method for assessment teams to use to develop their own socioeconomic projections of the factors of greatest local (or sectoral) interest was also provided, along with an exploratory approach using narratives to explore impacts.

A major issue with the scenarios prepared for NCA 2000 was that they were not widely used. Participants identified several obstacles to their use, including perceived lack of relevance (i.e., key information needs were not addressed by the model output); uncertainty based on the wide range of model projections between the two climate scenarios; and the limited use of historical data and sensitivity analyses (the other two modeling methods suggested for projecting future climate changes). The ecosystems/vegetation scenarios and socioeconomic scenarios were insufficiently linked to inform impact and adaptation analyses in an integrated fashion. The socioeconomic scenarios were not widely used because of perceived lack of relevance, provision late in the process, uncertainty about how to use them, and concerns about their relationship with the climate and ecosystems/ vegetation scenarios. In the few instances that a team did develop and apply context-specific

socioeconomic projections, the scenarios became overly complex, making them less plausible in hindsight.

In NCA 2009, scenarios were primarily used to provide context and illustration, rather than to stimulate analyses and assessments at the regional or sectoral level, as was the original (and only partially realized) intent in NCA 2000. For climate change information, the NCA 2009 used 16 models' simulations considered by the Phase 3 Coupled Model Intercomparison Project (CMIP3) conducted by the World Climate Research Programme's Working Group on Coupled Models (WCRP/WGCM) for the conterminous U.S. Alaskan projections were based on 14 models that best captured the present climate of the state. Caribbean and Pacific islands analyses used 15 models' simulations from the CMIP3 that were available at finer-scale resolutions. The model solutions were forced by two of the IPCC emissions scenarios (Special Report on Emissions Scenarios (SRES) A2 and B1 (Nakicenovic and Swart, 2000), spanning a range of high to low forcing, and for some applications a very high-emissions scenario (A1-"Fossil-Intensive" or A1FI) was also used. Based on CMIP3 results, the 2009 NCA offered broad interpretations and maps of the potential future regional implications of climate change for the U.S. Statistical downscaling was performed for a number of regions, and the results of this exercise informed several analyses and products within the assessment (e.g., "migrating states" maps, heat stress and mortality projections for selected cities around the country, trends in peak stream-flow timing for the West, Gulf Coast roads at risk from sea-level rise, and vegetation shifts in the Northeast). Unlike NCA 2000, NCA 2009 did not develop detailed socioeconomic scenarios for use in impacts and adaptation studies in the various regions and sectors.

NCA 2009 scenarios were used by members of the research community in preparing the published report, but since this assessment did not seek to engage stakeholders in regions or sectors, their utility for the purpose of stimulating analysis and deliberation among participants is unknown.

This experience points to a number of steps that could improve provision and application of scenarios for use in the NCA 2013 report and to develop scenario resources for use in ongoing distributed analyses and assessments. These include

- Build scenarios into the overall strategy for the Assessment (both regional and sectoral participatory processes and expert preparation of the 2013 report); ensure that the scenarios have credibility, salience, and legitimacy to both intermediate and end users; and provide clear guidelines and facilitation to help end users apply them.

- Balance centralized and decentralized scenario development in a manner that allows for some coordination across different activities and analyses, but also allows flexibility and adaptation of the framework and materials to be relevant to participants in sectors and localities across the country.

- Prepare and disseminate scenarios in a timely fashion to participants in the Assessment process.

- Ensure scenarios represent a wide range of future conditions to reflect uncertainty and to facilitate exploration of potential high-consequence, low-probability events in the tails of the distribution.

- Use the full range of information and methods available to develop quantitative scenarios and avoid the trap of relying on a limited set of information sources (e.g., only one or two models or downscaling techniques) for developing descriptions of future regional changes in climate and other conditions, and provide guidance to match the information, method, or model to the Assessment need.

- Develop tools and capacity that facilitate participatory use of scenarios by end users in the sustained NCA process.

- Take advantage of already completed model results and scenarios and develop new scenarios as needed to fill gaps.

It is unlikely that scenarios for NCA 2013 will be able to address all of these points, especially given limitations of time and resources, but concerted attention should be given to improving performance in at least one or two of these areas.

1.5 Needs for NCA 2013: Coordinating Modeling and Analyzing the Implications of Uncertainty for Decision Options

Scenarios provide a means of (1) establishing common bounding assumptions for modeling, research, and assessment, and (2) enabling decision makers to explore the implications of uncertainty about the future for their decisions. These different users and uses of scenarios create potentially conflicting demands for centrally provided data and information that ensure coordination across model types and a decentralized participatory strategy that can be adapted to conditions and uncertainties in different parts of the country. Based on the evaluation of prior experience with scenarios and the discussions at the workshop, the overall strategy for scenarios in the NCA has an opportunity to address both of these needs. Both can be accomplished by centrally providing a limited set of scenario data as well as technical guidelines and tools to enable the regional and sectoral teams participating in the Assessment to develop and apply decision-support scenarios that are nested within the centrally provided scenario data. This section of the chapter explains this opportunity and discusses options for the overall scenario strategy of the Assessment.

The NCA could provide sectoral and regional assessment teams with a set of integrated scenario materials that impart assumptions about potential future climate, environmental, and socioeconomic conditions at the scale of eight to ten regions of the country that the 2013 report is likely to include. For climate change, the NCA could prepare climate change trends and outlooks for major regions of the country describing what is known about the range of potential future conditions in approximately 25 and 100 years, drawing on information and data from global climate models, downscaling, observations and historical records, and recent research into climate processes known to be important in a given region. The climate change outlooks for each region would describe what is known about the temperature, precipitation, and other aspects of climate consistent with high and low levels of climate change for each region, based on an agreed set of global scenarios.

The socioeconomic and environmental information would also focus on trends at national and regional scales and focus on conditions that affect adaptive capacity and vulnerability (e.g., household, labor force, and sectoral composition of the economy). For 2035, existing model results would be drawn from the agencies that specialize in making these projections, and for 2100 less detailed projections could be provided from an integrated-assessment model. To provide historical context, 100-year U.S. historical time series of population, gross domestic product, and labor productivity could also be provided, showing how the future scenarios fit within the realized historical patterns. Focusing on larger, aggregate regions seems to be a good option both because of limits to predictability at finer spatial scales for the periods of interest, defined by the Global Change Research Act (25 and 100 years into the future), and because centrally-provided, long-term projections at a state or county level are not likely to be credible to most stakeholders. The NCA would be well advised to take advantage of data or projections already produced by agencies such as National Oceanographic and Atmospheric Administration (NOAA), the Department of the Interior, the Bureau of Economic Analysis, or the Bureau of the Census, and by scientific activities such as CMIP3 and the North American Regional Climate Change Assessment Program (NARCCAP).

The decentralized participatory strategy would use the climate, environmental, and socioeconomic information provided as background and have the sectoral and regional teams construct narratives for their region or sector that describe futures in which (1) they are able to adapt selected attributes or systems to the scenarios of change depicted in the climate, environmental, and socioeconomic scenarios, and (2) adaptive capacity is exceeded in some regards or situations, with consequences described in the scenarios (in the words of some workshop participants, the "things that keep them up at night"). Using the decentralized strategy to address choices and options confronting decision makers would increase its relevance. The sectoral and regional scenarios could include information on interactions of climate change with development objectives, assumptions about resources developed locally and those provided by the federal government, and factors important in explaining the success (or failure) of their adaptation efforts. Technical guidelines and facilitation would be required to foster consistent implementation of the strategy.

This overall approach has a number of potential benefits, including that the regional and sectoral

scenarios developed will have credibility, legitimacy, and salience; the scenarios will draw on detailed local knowledge of conditions and trends; the nesting of regional or sectoral scenarios in broad national scenarios depicts the situation that will likely confront decision makers (i.e., that the climate system and national processes will provide constraints and opportunities largely outside of their control in which they will need to operate); and the process of developing the scenarios will encourage discussion and examination of the interactions of climate change with development objectives that reflect values and desires of key groups within each region or sector. There are also a number of challenges to the strategy, including getting assessment teams to prioritize a few key attributes, systems or activities; defining "success" and "failure" of adaptation; and achieving consistency across regions and sectors. As the regional and sectoral scenarios are completed, the NCADAC could integrate insights from the information generated from the regional and sectoral assessments, possibly developing national-scale narratives.

This option for a scenario strategy integrates development of the regional or sectoral scenarios into the broader Assessment process and should be seen as part of the overall NCA strategy for regions and sectors. Effective implementation would depend on making use of information from previous assessments and addressing a number of specific issues including the range of forcing scenarios (and hence future climate change) that the Assessment should adopt; the specific variables and products to be included in climate, environmental, and socioeconomic scenarios; the content of technical guidelines and development of training and facilitation for assessment teams to apply the scenarios; an approach for synthesis of scenarios produced by sectoral and regional activities; and ideas for evaluation of the approach that is eventually adopted by the NCADAC.

1.6 Options and Next Steps for NCA 2013

Building on options explored in the scenarios white paper and points made during presentations and discussions at the workshop, a number of options and next steps are available to the NCADAC for scenarios that support the Assessment's objectives.

1.6.1 Need for an Accepted Strategy and Priorities

If scenarios are to be used by regional and sectoral assessment activities in the NCA, a clear strategy is needed for their use, and this strategy must be understood and accepted by users at the outset. The scenarios need to provide information relevant to the vulnerability and impact questions being asked, and any centralized materials provided to the sectors and regions must be delivered in understandable and usable forms early in the process of analysis and assessment. Generally, this means that scenarios based on new model results or research are not likely to be developed for the production of Assessment products that are on a tight time schedule, such as the 2013 report.

In a process as large and diverse as the NCA is expected to be, it is not possible to develop scenarios that will meet the needs of all the issues to be addressed by individual regional and sectoral assessments. The NCADAC would be advised to maintain realistic expectations and to prioritize what products are developed, keeping in mind the most important uncertainties in the key issues that the process is addressing. Decisions will also be needed regarding variables, time frames, spatial resolution, and other factors.

A number of options for scenarios from which priorities might be set for the NCA 2013 report (short-term – "S") or the sustained Assessment process (long-term – "L") identified in the white paper and discussions at the workshop include

- Climate outlooks for 25-year and century scales including information on the regional implications of high and low forcing scenarios (S),

- Historical climate analogs (S),

- Spatial climate analogs (e.g., transposed climate from one region to another) (S),

- Environmental scenarios of large-scale processes such as land-use change or sea-level rise (S),

- Simple bounding scenarios of socioeconomic conditions at aggregate regional scale (S),

- Socioeconomic narratives (L),

- Spatially-explicit quantitative scenarios (L),

- A family of scenarios of possible futures for systems besides the global climate, as essential components of multiple causation and stress assessments (L), and

- Scenario-based decision-support tools (L).

Other possible components of the strategy for scenarios include

- Technical guidelines for use of scenarios in regional or sectoral assessments and linkage to appropriate models (Earth system; integrated assessment; impacts, adaptation, and vulnerability; and environmental models), including
  - Templates, good practices, lessons learned, examples, and other guidance material for regional and sectoral groups (S), and
  - Methods and scenario-planning resources for the ongoing Assessment process (L);

- An inventory of already available scenarios and resources that can be incorporated into the NCA (S); and

- Capacity building, training, workforce development, and support for post-engagement follow-up (e.g., tools, Web manuals and technology, expert directories, communities of practice) (L).

1.6.2 Bounding Uncertainty

A key design question in establishing a framework of scenarios for an assessment is selecting the number of scenarios of anthropogenic forcing (and associated changes in climate) used and establishing a range for these conditions that accurately represents uncertainty in the state of science. This will be an important issue for the NCADAC to address early in their deliberations. How many scenarios should there be? With an odd number, the mid-range estimates are often perceived as the most likely, which is inaccurate and negates the purpose of having a set of scenarios that encourages examination of the implications of the distribution of possible outcomes. Particularly because of the tight time schedule anticipated for NCA 2013, there is likely to be value in using a small number of scenarios that bound uncertainty in setting some key assumptions for the more distributed process for creating scenarios for various purposes.

Regarding the range of uncertainty in forcing and climate, it is advisable to develop a range that permits focus on the low-probability, high-impact events. Failure to consider some scenarios and impacts that are extremely unlikely but consequential (e.g., extreme floods, epidemics, and other events) could leave decision makers unprepared to deal with future consequences that would be difficult to manage or respond to without advance planning. Decision makers often want to be prepared for even a worst-case scenario, and many decision processes have a lot of conservatism built into them to address such outcomes. To avoid being tagged as alarmist or speculative, there is a tendency in assessments, and even in some research, to move toward the middle of the range of possible outcomes. The NCA will need a transparent and defensible process for considering the range of futures to be studied in the Assessment, one that enables the Assessment to address extremes that have low probability, but high consequence, as is often done by the Department of Defense and other agencies to promote national preparedness.

1.6.3 Linkages across Spatial Scales and between Scenarios and Impacts

Scenario processes, elements, and outcomes can be linked across scales either strongly or weakly. A clear overall question for the scenarios for the NCA is deciding what the best degree of linkage is. A high degree of consistency across scales is desirable from a scientific modeling perspective, particularly as modeling focuses on increasing spatial and temporal resolution and tighter coupling across the issues associated with emissions, climate, and impacts and their interactions and feedbacks. In practice in an assessment, however, tight linkage is not always necessary or at least not the most important goal. Multi-scale scenarios (in which climate, economic, or others trends at the continental or higher scale serve as the broad context in which finer scale scenarios are developed) are important because different processes (including drivers), stakeholders, and decisions are important at different scales. Decision-support scenarios such as those that might be developed by a municipality or corporation to examine the implications of uncertainty in climate and other conditions for decisions about investments or infrastructure favor a loose coupling across scales because this allows more flexible scenario construction, promoting relevance and credibility in their decision-making process.

Experience to date indicates that linkages between scenarios and impact assessment research are often indirect and sometimes entirely missing. Much of the published climate change impact research does not start by specifying particular quantitative scenarios. Trying to ensure consistency regarding scenarios would mean excluding a majority of the relevant impact research. The NCA will need to consider a strategy for connecting the results of sensitivity studies of impacts (research that studies changes in crop production, water resources, or other systems resulting from specified arbitrary changes in temperature, precipitation, and other variables) with time-dependent, spatially-explicit scenarios. This issue will be a challenge for NCA 2013, and a key question for the NCADAC is whether a scenario strategy can be implemented that will strengthen the linkages with impacts research in the long run.

1.6.4 Visualization and Communication

It is important for the scenario process to plan how to communicate - challenging quantitative and technical material to sectoral and regional assessment teams as well as the broader audience of the Assessment. Scenarios rendered through visualizations can be a powerful approach. Giving first consideration to current conditions and plans for development, instead of climate change per se, can provide a more targeted and salient framing of the issues than simply starting with climate change scenarios. As a general rule, arraying findings from multiple scenarios is more informative (and believable) for risk perception and management than reporting results of a single scenario. Scenarios incorporated in assessment reports sometimes do not adequately address uncertainties, and this issue will also need to be addressed as discussed in several sections above.

1.6.5 Ad Hoc Group on Scenarios

The three chapters of this report help frame the main issues and conceptual challenges in scenario development for the NCA. However, it was never the intent of these foundational efforts to build a detailed strategy or actual scenarios for the NCA. How then to move forward? One option that has been raised would be for an ad hoc group on scenarios to address the specific issues and topics raised in this report as part of a more detailed scenario development and implementation strategy. If such a group is established, its focus could usefully include creating one or more options for scenarios in the short term as well as plans for making scenarios and scenario-planning tools both accessible and useable for the long-term, sustained Assessment activity.

Chapter 2:
Background White Paper on Scenarios for Assessing Our Climate Future – Issues and Methodological Perspectives

Richard Moss, Pacific Northwest National Laboratory; Nathan Engle, Pacific Northwest National Laboratory; John Hall, U.S. Department of Defense; Kathy Jacobs, Office of Science and Technology Policy; Robert Lempert, RAND Corporation; Linda Mearns, National Center for Atmospheric Research; Jerry Melillo, Marine Biological Laboratory; Phil Mote, Oregon State University; Sheila O'Brien, U.S. Global Change Research Program; Cynthia Rosenzweig, NASA Goddard Institute for Space Studies; Alex Ruane, NASA Goddard Institute for Space Studies; Stephen Sheppard, University of British Columbia; Bob Vallario, U.S. Department of Energy; Arnim Wiek, Arizona State University, and Tom Wilbanks, Oak Ridge National Laboratory

The authors thank workshop participants for their comments.

2.1 Introduction

This white paper was written as background for a workshop for the National Climate Assessment (NCA) that focused on the use and development of scenarios. The paper is included as a chapter in the report of the workshop because the authors and members of the organizing committee believe it conveys information of use to participants in the Assessment process, and the broader research and user communities that work with scenarios in climate science. The paper briefly defines key terms and establishes a conceptual framework for developing consistent scenarios across different end uses and spatial scales. It reviews uses of scenarios in past U.S. National Climate Assessments and identifies potential users of and needs for scenarios for both the report scheduled for release in 2013 and to support an ongoing, sustained Assessment process in sectors and regions around the country. Because scenarios prepared for the NCA will need to leverage existing research, the paper takes account of recent scientific advances and activities that could provide needed inputs. Finally, it considers potential approaches for providing methods, data, and other tools for Assessment participants.

We note that the term "scenarios" has many meanings. An important goal of the white paper (and portions of the workshop agenda) is pedagogical (i.e., to compare different meanings and uses of the term and make Assessment participants aware of the need to be explicit about types and uses of scenarios).

In climate change research, scenarios have been used to establish bounds for future climate conditions and resulting effects on human and natural systems, given a defined level of greenhouse gas emissions. This quasi-predictive use contrasts with the way decision analysts typically use scenarios (i.e., to consider how robust alternative decisions or strategies may be to variation in key aspects of the future that are uncertain).

As will be discussed more fully below, in climate change research and assessment, scenarios describe a range of aspects of the future, including major driving forces (both human activities and natural processes), changes in climate and related environmental conditions (e.g., sea level), and evolution of societal capability to respond to climate change. This wide range of scenarios is needed because the implications of climate change for the environment and society depend not only on changes in climate themselves, but also on human responses. This degree of breadth introduces a number of challenges for communication and research.

2.1.1 Definitions and Types of Scenarios

In this white paper, the term "scenarios" will be used to describe qualitative and quantitative information about different aspects of the future developed to investigate the potential consequences of climate change. There are a number of excellent general references on the use of scenarios in climate change research. This paper draws heavily on Parson et al. (2007), a review of scenarios prepared as one of the U.S. Climate Change Science Program (CCSP) Synthesis and Assessment Products. Drawing on this review and other references, this white paper classifies scenarios according to their content and the types of models or methods used to produce them. According to this typology, the major types of scenarios relevant to the NCA include emissions, climate, environmental, socioeconomic, and narrative.

Emissions scenarios are descriptions of potential future emissions to the atmosphere of greenhouse gases and other radiatively important gases and particles that are used to explore the implications of alternative energy and technology futures and provide inputs to climate models. Emissions scenarios are not forecasts or predictions. They focus on long-term (e.g., decades to centuries) trends in energy and land-use patterns, not short-term fluctuations. They are developed using integrated assessment models and are based on research into socioeconomic, environmental, and technological trends. Uncertainty in emissions scenarios results from the inherent uncertainty about future socioeconomic and technology conditions and differences in representations of processes and relationships across models, among other factors. Fisher et al. (2007) evaluate recent emissions scenario literature, and Weyant et al. (1996) provide an overview of integrated assessment modeling approaches.

Climate scenarios are plausible representations of future climate conditions (temperature, precipitation, and other factors) produced using a variety of techniques including scaling of observed climate, spatial and temporal analogs in which climates from other locations or periods are used as example future conditions, extrapolation and expert judgment, and mathematical climate and Earth system models. All of these techniques continue to play a useful role in development of scenarios, with the appropriate choice of method depending on the intended use of the scenario. Regional-scale climate scenarios and projection methods for impact and adaptation assessment are highly relevant for the NCA (Mearns *et al.*, 2001).

Environmental scenarios focus on changes in environmental conditions such as water availability and quality, sea-level rise (incorporating geological and climate drivers), land cover and use, and air quality. Climate change can drive changes in these factors, or scenarios can represent independently caused variations. The potential impact of climate change and the effectiveness of adaptation options cannot be understood without examining interactions of changes in climate, environmental conditions, and human responses (Carter *et al.*, 2001).

Socioeconomic scenarios for assessment of impacts, adaptation, and vulnerability project future demographic, economic, institutional, and other characteristics that are needed for different types of impact modeling and research. This information is crucial for evaluating the potential to be affected by changes in climate as well as for examining how different types of economic growth and social change affect the capacity to adapt to potential impacts. Many of the same socioeconomic factors that affect emissions also affect vulnerability and adaptive capacity and thus the underlying socioeconomic modeling must be coordinated. Nakicenovic *et al.* (2000) summarize socioeconomic driving forces. For a description of needs for socioeconomic scenarios and narratives, see a recent report of the U.S. National Research Council, National Academy of Sciences (NRC, 2010).

Narratives describe in qualitative form political, institutional, and other factors that influence future forcing, vulnerability, and responses. Narratives are useful because while some socioeconomic factors affecting emissions and vulnerability are modeled quantitatively, others are not effectively quantified. Narratives can be used as the basis for quantitative scenarios, as in the Intergovernmental Panel on Climate Change (IPCC) Special Report on Emissions Scenarios (SRES) (Nakicenovic and

Swart, 2000). They can also facilitate coordination across spatial scales and substantive domains (NRC, 2009, Zurek and Henrichs, 2007). More broadly, narratives are "stories" about the future that are developed strategically to lead decision makers (end users) to consider futures and potential responses that they might have otherwise neglected but that are nonetheless important. The use of scenarios as explicit decision-support tools contrasts with the ways in which narratives can be employed by researchers to coordinate studies across scales or sectors.

Other typologies of scenarios are also available and focus on audience, use, and other characteristics. Bradfield *et al.* (2006) categorized scenarios into three schools: intuitive logics (exemplified by the work of Rand, the Global Business Network, and Shell), in which a small number of diverse scenarios are crafted that help decision makers understand the most important drivers of their future and how best to respond); "La Prospective" or other backcasting methods (e.g., Godet, Berger) in which desirable futures are defined and the scenarios specify how these visions might be attained; and Probabilistic Modified Trends (e.g., Gordon, Helmer), which aim to add surprise to traditional forecasting methods. Another typology, proposed by van Notten *et al.* (2003), differentiates scenarios according to their goal (raising awareness or decision support); the process used to create them (interactive group sessions or a formal process employing quantified knowledge); and the scenario content (complex or simple).

2.1.2 Scenario Users

Two broad categories of users of scenarios are often distinguished: *intermediate users* (modelers and other members of the research community) and end users (decision makers, stakeholders, and others). This distinction is established here, briefly described, and further developed in subsequent sections of the white paper, especially Section 2.4, which focuses on potential scenario products for different sets of intermediate and end users.

Intermediate users: In previous assessments (NCA 2000 and NCA 2009), scenarios were developed primarily for intermediate scientific users to provide information from one area of research to another (Figure 2.1). This effort was essential for researching and writing the assessment reports themselves and is described in more detail in Section 2.3. The need to

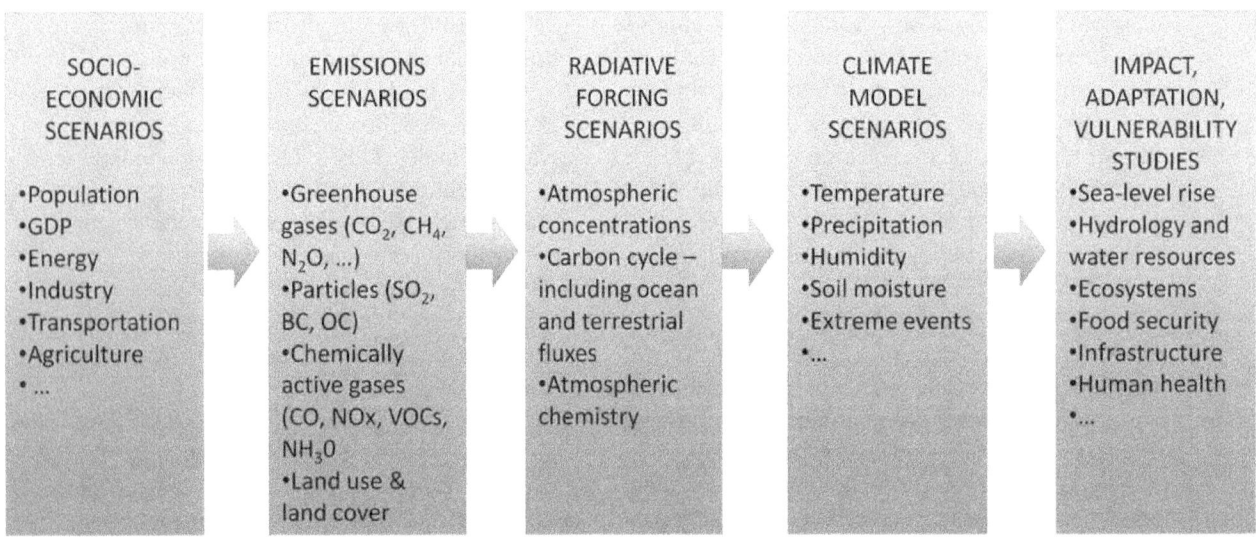

SOCIO-ECONOMIC SCENARIOS

- Population
- GDP
- Energy
- Industry
- Transportation
- Agriculture
- ...

EMISSIONS SCENARIOS

- Greenhouse gases (CO_2, CH_4, N_2O, ...)
- Particles (SO_2, BC, OC)
- Chemically active gases (CO, NOx, VOCs, NH_3O
- Land use & land cover

RADIATIVE FORCING SCENARIOS

- Atmospheric concentrations
- Carbon cycle – including ocean and terrestrial fluxes
- Atmospheric chemistry

CLIMATE MODEL SCENARIOS

- Temperature
- Precipitation
- Humidity
- Soil moisture
- Extreme events
- ...

IMPACT, ADAPTATION, VULNERABILITY STUDIES

- Sea-level rise
- Hydrology and water resources
- Ecosystems
- Food security
- Infrastructure
- Human health
- ...

Figure 2.1. Typical sequential hand off of information across scientific disciplines using a range of scenario types (Moss *et al.*, 2010).

coordinate and integrate different types of analyses with scenarios will be important for preparing the NCA 2013 report.

End users: this category of users is very diverse and includes elected officials, resource managers, land-use or urban planners, entrepreneurs, analysts and executives in the private sector, nongovernmental organizations, citizens, and many others who are the ultimate audience of assessments or who wish to assess the need to take account of climate change in their future activities and plans. By design, the scenarios workshop did not involve end users but instead relied on inputs from other workshops supporting the NCA that did include them (e.g., see Section 2.4.1 below for a summary of needs identified at an NCA workshop on sectors and regions). The workshop did include individuals who work in "boundary" or "bridging" organizations that interact with a variety of end users by interpreting and assisting with application of scenarios and other research-based methods of decision support.

In the long-term NCA process, greater emphasis will be given to providing tools and building capacity to support assessment and deliberation at local, state, and regional levels. To facilitate these distributed activities and future NCA reports, it will be necessary to develop methodologies that can be adapted and applied across a range of regions and sectors of the U.S. For these uses, greater attention will be paid to developing participatory scenario processes that enable end users and local analysts to consider context-specific decisions

throughout a range of climate, socioeconomic, and environmental scenarios. The primary audience for these scenarios and scenario products will be those regional and local decision makers who are developing climate action plans or simply want to reflect upon ways in which climate change may affect their interests. The purpose of scenarios for these individuals is to improve decision making by helping practitioners consider alternative climate futures and impacts, identify key vulnerabilities, and gauge capacity to adapt or mitigate, among others. For example, managers might use the NCA environmental and socioeconomic scenarios to help draft a forest management plan for a given region, or use narratives to perform long-term visioning and planning for their community.

An underlying issue related to the use of scenarios in decision making is whether probabilities can be usefully associated with scenarios (e.g., Desai and Hulme, 2004; Grubler and Nakicenovic, 2001; Hall, 2007; Katz, 2002; Knutti *et al.*, 2005; Nakicenovic *et al.*, 2000; Pittock *et al.*, 2001; Schnieder, 2001). The motivation for providing probabilistic representations of scenarios is that without quantification of relative likelihoods, decision makers will have insufficient information upon which to base decisions or will develop their own assessments of relative likelihood that depart from the best judgment of experts. A number of concerns have been raised, however, including that the resulting estimates may overstate existing knowledge of probabilities of different potential futures, under-represent uncertainty, or that even attempting to attach probabilities to scenarios conflicts with their proper use in decision-making contexts. Another concern regarding use of scenarios is that users can develop overconfidence in them. Any scenario or set of scenarios will represent only a small fraction of possible futures, yet when people interpret them, they can believe that they represent all or the most important or likely possibilities. Overconfidence is particularly likely without the explicit assignment of probabilities to specific scenarios. However, as discussed above, the assignment of probabilities is controversial. It is thus essential that scenarios prepared for use in the NCA be accompanied by clear guidance on their interpretation, uses, and limits.

2.2 Overview of Strategy for NCA 2013 and Ongoing Distributed Climate Assessments

2.2.1 Vision and Goals

Scientific assessments serve as progress reports by identifying advances in the underlying science, providing critical analysis of issues, highlighting important findings and key unknowns that can improve policy choices, and guiding decision making related to climate change. The approach that is envisioned for the NCA 2013 is a comprehensive assessment of climate change, impacts, vulnerabilities and adaptations, within a context of how communities and the Nation as a whole work to create sustainable and environmentally sound development paths.

The vision for the NCA 2013 is to establish a continuing, inclusive national process that
- Synthesizes relevant science and information;

- Increases understanding of what is known and not known;

- Identifies needs for information related to preparing for climate variability and change and reducing climate impacts and vulnerability;

- Evaluates progress of adaptation and mitigation activities;

- Informs science priorities;

- Builds assessment capacity in regions and sectors; and

- Builds societal understanding and skilled use of assessment findings.

2.2.2 Mandate and Focus

The mandate for the Assessment is contained in the Global Change Research Act (GCRA) of 1990.[1] Section 106 of the Act specifies that a "Scientific Assessment" must be prepared not less frequently than every four years and delivered to the President and Congress. This assessment must

- Integrate, evaluate, and interpret the findings of the Global Change Research Program, and discuss the scientific uncertainties associated with such findings;

- Analyze the effects of global change on the natural environment, agriculture, energy production and use, land and water resources, transportation, human health and welfare, human social systems, and biological diversity; and

- Analyze current trends in global change, both human-induced and natural, and project major trends for the subsequent 25 to 100 years.

This last requirement to analyze trends into the future requires the use of physical models at various scales, but also the ability to build scenarios that help describe and analyze future conditions where changes in climate are only one of a myriad of changing variables.

Although at the time of the workshop, the definition of regions to be used in NCA 2013 is still very much in flux, it has been noted that the ability to deploy information on the Web would significantly relieve the pressure on how to define the boundaries. If the Assessment can "nest" information within a number of national, regional, and local scales, the exact boundaries of the regions become much less important. That said, at the recent regional and sectoral workshop many participants felt that regions roughly analogous to those used in the 2009 report would be desirable, with adjustments to use state boundaries wherever possible. There is a strong desire for both understanding regional climatology and having the capacity to project conditions at the regional level, and at multiple timescales, including seasonal to inter-annual, decadal, and 50–100 years. The need to understand change in both a transient and endpoint

---

[1] http://www.gcrio.org/gcact1990.html

Suggested Assessment Structure

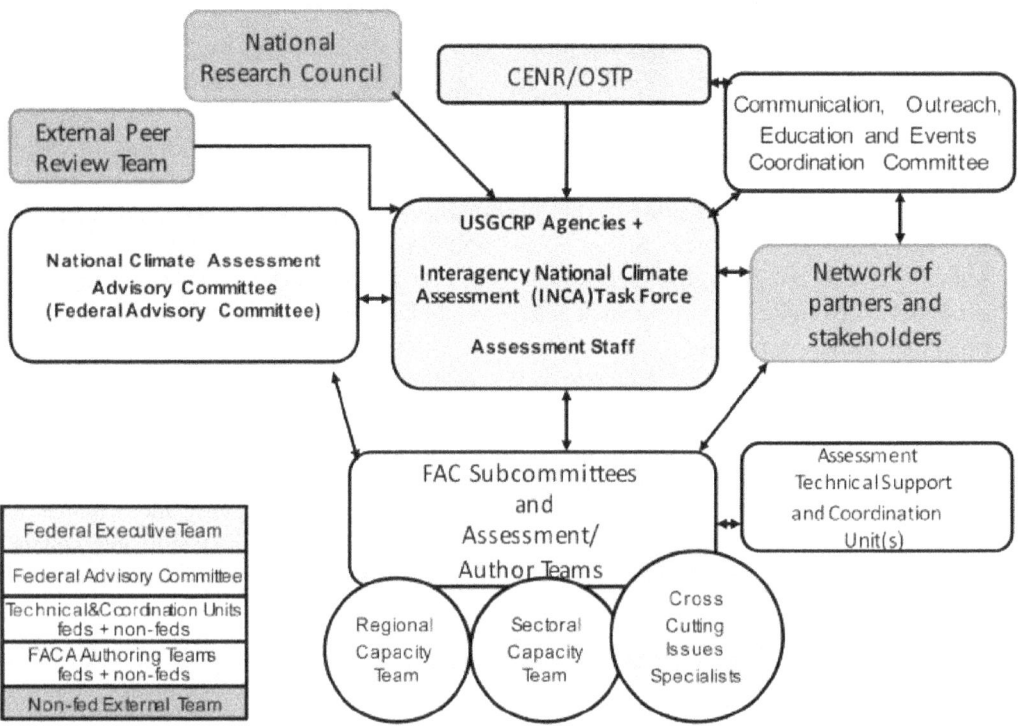

Figure 2.2. Suggested Assessment structure.

framework was also noted. Finally, the significant focus on engagement and communications raises special challenges for the intermediate user groups, who may be asked to help build coherent storylines for the future at the regional scale.

In addition to preparing the 2013 report, the NCA seeks to build distributed national capacity to assess the implications of climate and global change both inside and beyond the federal government. This ongoing process will draw upon the work of stakeholders and scientists across the country. Assessment activities will result in the capacity to do ongoing assessments of vulnerability to climate stressors, observe and project impacts of climate change within regions and sectors, develop consistent indicators of progress in reducing vulnerability, and allow for the production of a set of reports and Web-based products that are useful for decision making at multiple levels.

2.2.3 Process and Implications for Delivery of Scenarios

Overall direction for the NCA will be provided by the National Climate Assessment Development and Advisory Committee (NCADAC) to be constituted under the Federal Advisory Committee Act (FACA) by the Department of Commerce. The NCADAC will be charged with integrating and evaluating the findings of the USGCRP and balancing scientific, engineering, educational, legal, and policy expertise. The roles of a variety of organizations in preparing, reviewing, resourcing, and providing oversight for the NCA are illustrated in Figure 2.2.

Major milestones in the work plan for the Assessment include completion of a series of methodological workshops; completion of regional and sectoral workshops; completion of a review draft; and completion of the report to the President and Congress in 2013. Socioeconomic and climate data and scenarios should be provided to the regional and sectoral teams by the middle of 2011 to maximize the utility of that information to the Assessment process. An important consideration in the process is ensuring adequate opportunity for peer review and public comment on the draft Assessment report before its completion.

2.3 Past Uses of Scenarios in NCA 2000 and NCA 2009

A range of scenarios were developed and provided for both the 2000 and 2009 National Climate Assessments, which were very different processes in character and extent (NAST, 2001; Karl *et al.*, 2009). The NCA 2000 included attention to establishing an ongoing assessment process and produced a concise overview report, foundational reports for eight mega-regions (and most sub-regions, including underlying technical reports), five sectors, and a report on native peoples and homelands. Unfortunately, support to maintain the ongoing assessment process was not sustained (CCSP, 2003). The NCA 2009 report summarized information contained in 21 "Synthesis and Assessment Products" produced by the research program when it was known as the "Climate Change Science Program" from 2002–2009.

This section of the white paper summarizes the types of scenarios that were provided and developed for each report and how the scenarios were actually applied.

2.3.1 Climate Change Impacts on the United States: The Potential Consequences of Climate Variability and Change (NCA 2000)

Three basic categories of scenarios were developed and used for the 2000 assessment: climate, ecosystem/vegetation, and socioeconomic (MacCracken *et al.*, 2001; Melillo *et al.*, 2001; Parson *et al.*, 2001). The sections below provide a brief review of each of the categories, followed by preliminary lessons and questions that might inform development and use of scenarios for NCA 2013 and the ongoing Assessment process.

2.3.1.1 Climate scenarios
To ensure use of up-to-date results and to promote consistency across the broad number of research teams participating in assessment, the National Assessment Synthesis Team (NAST) developed a set of guidelines to identify simulations to be considered for use by the regional and sectoral teams of the NCA 2000. For a variety of reasons discussed by MacCracken *et al.* (2001), two model simulations—one developed by the Canadian Centre for Climate Modelling and Analysis (CGCM1) and the other by Hadley Centre for Climate Prediction and Research of the Meteorological Office of the United Kingdom

(HadCM2)—were recommended, both forced with the IPCC IS92a emissions baseline scenario. The full range of variables relevant for analysis of impacts was available through the modeling teams. The climate scenarios chapter of the assessment report provided a detailed analysis of the simulation results, focusing on a select set of variables and processes including temperature, precipitation, soil moisture, sea ice and level, extreme events, etc.

Authors used these two model simulation results, but several of the regions and sectors went beyond the two suggested global climate models and used a broader range of models and projections. The use of additional results derived from global climate models within some of the regions and sectors stemmed from a variety of factors, including the perceived lack of fit between the Hadley and Canadian models for a given region, greater capacity and financial resources that allowed more comprehensive model exploration, and a mismatch of timing between when modeling studies were commissioned within the regions and sectors and when the NAST suggested which global model results to use. In addition, some used various data sets and downscaling techniques to interpolate data on finer spatial or temporal scales (e.g., the Vegetation/Ecosystem Modeling and Analysis Project (VEMAP) for ecosystem responses, mesoscale models guided by the global climate model output as boundary conditions, and statistical downscaling based on local climate data).

Several of the major concerns and challenges with the climate scenarios identified by Morgan *et al.* (2005), MacCracken (2000), and in research for this white paper included perceived lack of relevance of the two models in some of the regions (i.e., the region's questions were not really answered by the model output); uncertainty caused by the wide range of model projections between the two climate scenarios; and the limited use of historical data and sensitivity analyses (the other two modeling methods suggested for projecting future climate changes).

2.3.1.2 Ecosystems/vegetation scenarios
The VEMAP project generated future ecosystems scenarios for the conterminous U.S. The VEMAP outputs were based on the two Hadley and Canadian model simulations, and the assessment groups used the scenarios to assist in sensitivity analyses. Outputs were based on biogeochemistry models in the near term (2025-2034) and

biogeography models in the longer term (2090-2099). Application of vegetation and ecosystem scenarios varied across regions and sectors, with some using an historical climate data set developed for use with VEMAP to provide gridded monthly averages for key variables, and others reviewing literature and soliciting expert and stakeholder input to understand how ecosystems would respond to various climate changes.

The major concerns and challenges included the lack of comprehensive use across all regions and relevant sectors and an insufficient linkage between the ecosystems/vegetation scenarios and socioeconomic scenarios to inform impact and adaptation analyses. These concerns and their implications for the modeling strategy in the NCA 2013 were addressed more fully in a subsequent workshop on models (December 8–10, 2010).

2.3.1.3 Socioeconomic scenarios

As mentioned above, the socioeconomic and emissions scenarios used to force the global climate models corresponded to the IPCC IS92a scenario, a mid-range "business as usual" emissions scenario with middle-of-the-road socioeconomic assumptions with respect to demographic, economic, and other conditions. Socioeconomic scenarios were also explicitly developed to provide context for evaluation of impacts, vulnerabilities, and adaptations. This involved both a centralized, or top-down track, and a decentralized, or bottom-up track.

The centralized track focused on providing county-level projections of a few key variables relevant to all regions and sectors to 2030 and aggregate national-scale projections to 2100. In the short term (2030), three projections (high, middle, and low for all variables) depended upon varying assumptions of fertility, mortality, migration, labor-force participation, and productivity by age group, based on data from sources such as the Census Bureau. The projections were run using a commercial regional economic growth model, provided by NPA Data Services, which calculated annual population projections by sex and five-year age cohort for each state, county, and metropolitan area. Consistency was not established between the socioeconomic forcing scenarios (which used the IS92a assumptions) and these detailed 2030 projections (which were not based on IS92a assumptions). The longer-term socioeconomic projections to 2100 were only made available at the aggregate

national level. These three longer-term scenarios were developed with an integrated assessment model and were intended to be consistent with three of the IPCC SRES scenarios. Evaluation of differences in the socioeconomic scenarios that resulted from the use of different assumptions and sources (IS92a, SRES, and Census Bureau) was not performed. A question to be explored in developing socioeconomic scenarios for NCA 2013 is the degree of consistency needed across scenario components.

The decentralized track provided a common method for assessment teams to develop their own socioeconomic projections of factors of greatest local (or sectoral) interest beyond the three variables projected in the centralized track. Also, within the decentralized track, in an exploratory approach using narratives, assessment teams were encouraged to walk through plausible socioeconomic conditions that might lead to a range of impacts, scouting for possible vulnerabilities and opportunities that might escape notice in a more conventionally structured inquiry.

Major concerns and challenges with the socioeconomic scenarios pertained mainly to their lack of use and questions about their relationship with the climate and ecosystems/vegetation scenarios. Assessment teams rarely used the centralized track, and utilized the decentralized track and exploratory approach even less. When socioeconomic scenarios were used, quantitative projections were prioritized over constructing storylines of alternative socioeconomic futures. In the few instances that a team did consider the context-specific variables, the scenarios became overly complex, making them less plausible in hindsight. When neither quantitative projections nor qualitative narratives were used, literature reviews, expert judgment, and case studies were used. As Morgan *et al.* (2005) point out, the majority of assessment participants surveyed after the NCA 2000 suggested that the social and economic impacts should be handled differently in future iterations of the assessment, albeit with little agreement on how to do so.

2.3.2 Global Change Impacts in the United States (NCA 2009)

In NCA 2009, scenarios were primarily used to provide context and illustration, rather than to stimulate analyses and assessments at the regional or sectoral level, as was the original (and only partially realized) intent in NCA 2000. A shortened time frame for producing this report (approximately 13 months) limited the opportunity for engagement of regional and sectoral stakeholders. Reliance on conclusions from the CCSP Synthesis and Assessment Products and the IPCC Fourth Assessment Report were also among the reasons for the approach taken.

For climate change information, the 2009 assessment used 16 models' simulations from the CMIP3 for the conterminous U.S. For Alaska, projections were based on 14 models that best captured the present climate of the state. Caribbean and Pacific islands analyses used 15 models' simulations from the CMIP3 that were available at finer-scale resolutions. The runs were forced by SRES A2 and B1 emissions scenarios, and for some applications a high-emissions scenario (A1-"Fossil-Intensive" or A1FI) was also used. Based on CMIP3 runs, the 2009 NCA offered broad interpretations and maps of the potential future regional implications of climate change for the U.S. Downscaling was performed for a number of regions, and the results of this exercise informed a number of analyses and products within the assessment (e.g., "migrating states" maps, heat stress and mortality projections for selected cities around the country, trends in peak stream flow timing for the West, Gulf Coast roads at risk from sea-level rise, and vegetation shifts in the Northeast, to name a few).

Unlike the NCA 2000, the NCA 2009 did not develop detailed socioeconomic scenarios for use in impacts and adaptation studies in the various regions and sectors.

2.3.3 Some Implications for Future NCA Scenarios

Published reviews or previous assessments (e.g., MacCracken, 2000; Morgan et al., 2005) and research for this white paper point to six key issues that should be addressed to improve provision and application of scenarios for use in the NCA 2013 report and ongoing distributed analyses and assessments:

1. Being clear about the types of scenarios (and relationships between different types) and information that is needed and will actually be used, which will be a function of the credibility, salience, and legitimacy of these materials to both intermediate and end users (as well as other issues below, especially including making

scenarios available early enough in the process to be useful);

2. Balancing centralized and decentralized scenario development in a manner that allows for coordinated guidance but also flexibility and adaptive learning on the part of participants in sectors and localities across the country;

3. Making scenarios available in a timely fashion to participants in the Assessment process;

4. Improving characterization and communication of uncertainty in scenarios used in the Assessment process, which is partly a function of relying on many sources of information (not just one or two models) for developing descriptions of future regional changes in climate and other conditions;

5. Developing tools and capacity that facilitate participatory use of scenarios by end users in the sustained NCA process; and

6. Taking advantage of already constructed scenarios and literature reviews and conducting new scenario analyses, storylines, case studies, and research as needed to fill gaps.

Elaboration follows regarding several of these points.

2.3.3.1 Coupling scenario types

Users seemed to have difficulty relating climate, ecosystems, and socioeconomic analyses and the interactions between them within each of the regions and sectors. There was some coupling of climate-ecosystems and physical (hydrology) models but little coupling of climate and socioeconomic models; for example, integrated assessment models, which couple all three domains, were rarely used. It is crucial to address relationships and consistency across different types of scenarios. Collaboration across distinct research and user communities engaged in scenario development and application is improving (Moss et al., 2010), but there are still limits to the extent to which absolute consistency can be established across emissions, climate, ecosystem, and socioeconomic scenarios. A clear explanation of the degree of coupling across these domains needs to be incorporated into explanatory materials that accompany the scenarios.

2.3.3.2 Balancing centralized and participatory scenario processes

In previous assessments, the decentralized and participatory approach to scenario development was not well coordinated with centralized guidance regarding scenarios. For the next report, it will be very important to be clear about the balance between the use of centrally-provided scenarios and regional or sectoral initiatives to define scenarios. At one end of the spectrum, the NCA could provide basic tutorials and guidance on Assessment objectives and methods and strategies for thinking about the future and leave it up to regional and sectoral assessment teams to develop their own scenarios. At the other, the NCA could attempt to require use of centrally-provided data and narratives. A key issue is maintaining comparability but allowing groups latitude to develop scenarios that have credibility and salience to key issues they identify.

2.3.3.3 Stakeholder engagement with scenarios

For the most part (and with some exceptions in different regions and sectors) stakeholder involvement in the scenario process for previous assessments was mainly at the beginning (framing) and conclusion (reviewing analyses for validity or simply receiving the report), but rarely in the scenario development and analysis. Most involvement centered on identification of key issue areas of concern to evaluate under future climates. Fewer analyses focused on formally asking stakeholders about perceived vulnerabilities and adaptation strategies, and even fewer included stakeholders in envisioning alternative futures. Improving engagement of stakeholders in other aspects of the Assessment process will be crucial for building a sustained, ongoing process.

2.3.3.4 Supporting assessment of mitigation as well as adaptation responses

Mitigation was not considered part of the previous assessments, but post-assessment evaluations suggested that it should have been. This is important for providing resources for localities interested in assessing the full range of responses and developing climate change action plans, which need to be based on inventories and projections of human and natural emissions sources, among other factors.

2.4 Products Needed for NCA 2013

In Section 2.1.2, the white paper differentiated "intermediate users" and "end users." These are, of course, general categories and within each, there are a number of specialized applications and needs that can be identified. For the purposes of discussion, however, the distinction is helpful in

identifying broad sets of scenarios and scenario-based products that could be useful (and used). These include (1) scenarios for intermediate users, especially to support and coordinate modeling and synthesis; and (2) scenarios and related tools intended to inform or support participatory processes and consideration of the implications of climate change in a range of decision and deliberative settings. There is some overlap in these two sets of needs, but there are also important tensions. For example, the former set of needs would benefit from consistent scenarios, whereas the latter set would benefit from a diversity of scenarios that take into account potential surprises (EEA, 2009). Thus, a difficult challenge for the NCA will be meeting the needs of both sets of users.

2.4.1 Needs Identified in the NCA Workshop on Planning Regional and Sectoral Assessments

A workshop on regional and sectoral assessments held in November 2010 involved stakeholders and researchers in identifying information needs and options for conducting these assessments. Many participants expressed support for using regions roughly analogous to those used in the 2009 report, with adjustments to follow state boundaries wherever possible. There were some suggestions for new regions, such as the Arctic. A strong desire was expressed for both understanding regional climatology and having the capacity to project conditions at the regional level at multiple timescales, including seasonal to inter-annual, decadal, and 50–100 years. Breakout groups identified a large number of potential sectors for consideration. Many participants sought increased emphasis on certain topics, such as the oceans, vulnerable communities, and societal responses to climate change. A dominant theme expressed at the workshop was the importance of framing climate change within a multiple stressor context.

One of the most important insights was the need to focus on crosscutting themes that integrate regional and sectoral issues and increase the applicability and usefulness of the Assessment process and products. For example, several participants identified the nexus of water, energy, and agriculture in the Southwest, the unique challenges facing urban areas (e.g., transportation, infrastructure, and public health), and oceans as important crosscutting topics for the NCA. It was also strongly suggested that in addition to emphasizing analyses across regions and sectors, it would be helpful to have deliberate overlap and interaction between regional and sectoral author teams and chapters.

Which regional definitions, sectors and crosscutting topics are emphasized in the final outline approved for the NCA 2013 report will have implications for the scenarios needed for the Assessment. Several overarching messages emerged from the workshop regarding potential needs for and uses of scenarios. These insights are related below to the needs of intermediate and end users of the NCA.

2.4.1.1 Intermediate users
Insights from the workshop include the need for
* Explicitly discussing the modeling metrics and uncertainties that are incorporated into the various scenarios, and how the models perform;

* Considering other ongoing assessment activities occurring within states and international contexts that might provide useful knowledge for how to guide the scenario development and application process (e.g., consideration of the climate atlas that will be produced for IPCC AR5 and how the scenarios and scenario products will connect with or build from this atlas approach, as well as data sets, tools, and scenario development processes that have been constructed for various state assessments that might help guide the national process and avoid "reinventing the wheel");

* Establishing some level of scenario consistency (e.g., a suite of climate and socioeconomic scenarios used to inform emissions, impacts, vulnerability, and adaptation scenarios), particularly at higher spatial scales, with flexibility to capture context-specific nuances at regional and local scales; and

* Evaluating early in the Assessment process whether intermediate user demand exists for very fine-scale projections for all regions and sectors (e.g., socioeconomic projections at the county level); and when these projections are demanded, providing clear centralized guidance for how or when to use them.

2.4.1.2 End users
Insights from the workshop include the need for
* Making the report itself more accessible and illustrative for the end user to see the bigger picture related to the synergies, tradeoffs,

and maladaptations associated with impacts, vulnerabilities, adaptation, and mitigation (perhaps by structuring the report to reflect the scenario development process itself—walking through a handful of important sectors, one-by-one, starting with narratives and storylines, and carrying these through the rest of the report); and

* Making the products useable and available online, through such means as GIS files and decision-support tools.

2.4.2 Options for Scenarios and Related Products
This section of the white paper identifies four broad potential sets of scenario tools or products that could be developed to meet the needs of both intermediate and end users:
* Socioeconomic narratives (qualitative descriptions of the future) and scenarios (related quantification) to explore issues in mitigation and adaptation;

* Climate "outlooks" describe what is known about the evolution of climate variability and change at regional scales, based on expert opinions and drawing on a range of model outputs, observational records, and process research;

* Quantitative scenarios of climate change, changes in environmental conditions (e.g., land use, sea level, water availability and quality, and air quality), and socioeconomic conditions (mentioned above); and

* Scenario-based decision-support tools such as visualization, simulation, gaming, decision theater, and other interactive approaches for relating potential climate and socioeconomic changes to stakeholder-driven decision processes.

To the greatest extent possible, the NCA will have to make use of already developed scenarios and data sets, and coordinate with other organizations and activities to jointly develop scenarios that can serve multiple purposes. The white paper includes information on ongoing activities in the IPCC, other organizations, and the research community that could provide sources of data and scenarios.

This leveraging approach will contribute to more timely delivery of scenarios and related products and make effective use of resources in the research community.

2.4.2.1 Socioeconomic narratives and scenarios

It is now widely recognized that vulnerabilities to climate change depend on more than altered patterns of precipitation, temperature, or extreme events. They also depend on where people are and where they are going (demography), what they are doing (economic patterns and changes), how they govern commerce and mobilize for action (institutions), what cultural values and social constraints exist, and what their tools are for coping (e.g., technologies, planning, and social networks). Without being able to think systematically about the future evolution of these socioeconomic conditions it is difficult to assess what future climate changes would mean for regions, sectors, and societies, especially in the longer term. Moreover, without narratives of such dimensions of the future as starting points, it is difficult to create internally consistent scenarios of driving forces for projections of greenhouse gas emissions.

As mentioned in the definitions section of the white paper, narratives are qualitative descriptions of political, economic, institutional, cultural, and other factors that influence aspects of the future. They are useful as a foundation for quantitative scenarios and to consider the effects of factors such as institutional arrangements (e.g., laws and organizations) that cannot be quantified. Historical and analytical approaches can be used to systematically develop narratives that are rigorous and research-based to explore the evolution of important environmental and socioeconomic processes. Narratives can be used to convey the overall logic of a set of scenarios to a variety of audiences (in this sense, they are sometimes referred to as "storylines"), and can be used as the basis for quantitative scenarios, as in the IPCC SRES (Nakicenovic et al., 2000). They can also facilitate coordination across spatial scales and substantive domains (NRC, 2009; Zurek and Henrichs, 2007). Some narratives are normative and explicitly explore value-based desired end points such as "sustainable futures" while others are primarily descriptive and explore the implications of different trends and choices as they extend into the future.

Developing socioeconomic scenarios to accompany climate change scenarios over periods of many decades has, however, been difficult—partly because changes in human societies over long periods can be complex and profound. As a result, the socioeconomic sciences generally avoid projections that extend beyond a few decades.

Narratives could be developed for the NCA to frame assumptions about international and national developments that have consequences for vulnerability, resilience, adaptation, and mitigation in different regions, sectors, or jurisdictions of the country. For example, narratives could be developed to provide a framework for analysis of the implications of different approaches to national climate policy or different levels or types of economic growth across jurisdictions, regions, or sectors. Precedents for such cross-scale narrative frameworks that link adaptation and mitigation have been developed and, to some extent, tested (e.g., Sheppard et al., 2011). It is crucial, however, to be realistic about what can be produced on different timescales. For the 2013 NCA report, it may only be possible to provide already available narrative materials that might help to frame impact assessments, such as those developed by the National Park Service for the purposes of considering the implications of changes in climate and socioeconomic conditions on specific parks and facilities. Another option is to ask sectoral and regional assessment teams to develop narratives that focus on their priority issues or attributes. For this option to be viable, it would be necessary to provide technical guidelines and facilitation. For the longer-term sustained National Climate Assessment infrastructure, it may be possible to develop a study to produce socioeconomic narratives to accompany climate change projections as a basis for assessing regional, sectoral, and societal impacts in the U.S. (perhaps under the auspices of USGCRP, including multi-agency and stakeholder consultations).

There are a number of recent and ongoing activities on which to build. To explore what might be possible in socioeconomic scenario development, the National Academy of Sciences, National Research Council (NRC) organized an international workshop in Washington, D.C. in February 2010 that brought together a wide range of socioeconomic scientists and climate change modelers and analysts (NRC, 2010). Besides considering prospects for relatively long-term quantitative projections of such variables as demographic and economic change, the workshop considered such alternative approaches as qualitative socioeconomic narratives, as in the case of the Millennium Ecosystem Assessment.

IPCC Working Groups II and III are developing guidance to chapter authors about socioeconomic contexts for their assessments of both impact and mitigation prospects. To support this process and catalyze development of socioeconomic scenarios by the research community, an IPCC Expert Meeting on Socioeconomic Scenarios was convened November 1–3, 2010, in Berlin. The workshop continued the exploration of socioeconomic narrative and scenario development started at the NRC workshop. It included, as starting points, two white papers proposing different frameworks for developing socioeconomic narratives (Kriegler *et al.*, 2010; van Vuuren *et al.*, 2010). The meeting produced an agreement to develop a small number of Shared Socioeconomic Pathways (SSPs), associated with supporting quantitative scenarios where possible, related to major non-climatic driving forces for development paths. These SSPs will then be matched with climate scenarios to support impacts, adaptation, and vulnerability assessments.

2.4.2.2 Climate outlooks

The proposal for producing expert-judgment-based descriptions of the possible evolution of climate conditions at the scale of eight to ten larger regions of the U.S. grows from the observation that there are many sources of information on the range of possible climate futures. While Atmosphere-Ocean General Circulation Models (AOGCMS) remain the primary source of information on the range of possible climate futures, there is an increasing array of Regional Climate Models (RCMs) and downscaling approaches that also provide insights. An important issue to consider with use of global climate models is that they are constructed to obtain a best estimate of the likeliest climate sensitivity, not the range of sensitivity. Thus, these models may provide results that are too narrow. Processes for intercomparison of global and regional climate models are underway, and methods are continuing to mature. In addition, knowledge of processes shaping regional climate change is also growing and can add value to understanding how regional climates may evolve, especially when these processes are not yet incorporated adequately into AOGCMs or RCMs, for example, because they occur at scales finer than the computational

resolutions of the models. Finally, observations of recent conditions and changes in climate are an additional valuable source of information.

No model-based method is available for integrating these four sources of information (global climate models, downscaling approaches, process knowledge, and observations), hence it is proposed that for the NCA, expert panels would draw information together focusing on both a common set of variables across regions (starting with agreed temperature and precipitation variables) as well as on topics or processes of special interest within each region. Including representatives of regionally-significant user communities or boundary organizations in the process would help ensure that the information produced addresses key questions and climate features. This approach would facilitate nuanced expert assessment of key processes and features important to climate in each region. The outlooks would be presented as opening sections of each regional (and if relevant, sectoral) chapter of the NCA 2013 report. In addition, if possible, the climate outlooks should provide insights for regional users into appropriate model runs and scenarios for analysis within each region. This could be a particularly valuable function of the outlooks if, as seems likely, some earlier climate model results will need to be used in some regional or sectoral analyses because new model runs are not yet available. The outlooks would provide a means for the expert community to provide information to users on differences and the implications of new information just being made available for earlier sets of scenarios. Uncertainty is a key issue and the type of uncertainty changes over time and over spatial scale. Natural variability dominates on short timescales (less than 10 years), but inter-model and emissions uncertainty become more important further into the future (Hawkins and Sutton, 2010a, b). The construction of scenarios—and the choices about models to include and how to include them, in order to quantify uncertainty—therefore depends partly on the timescale of relevance. Information to guide users through this thicket of issues would be extremely valuable for ongoing distributed Assessment activities.

The outlooks should include maps and figures. A number of effective presentations based on downscaling results were used in NCA 2009, and in addition there are relatively simple graphical approaches that can be used to portray the spread of model results for key variables and to compare

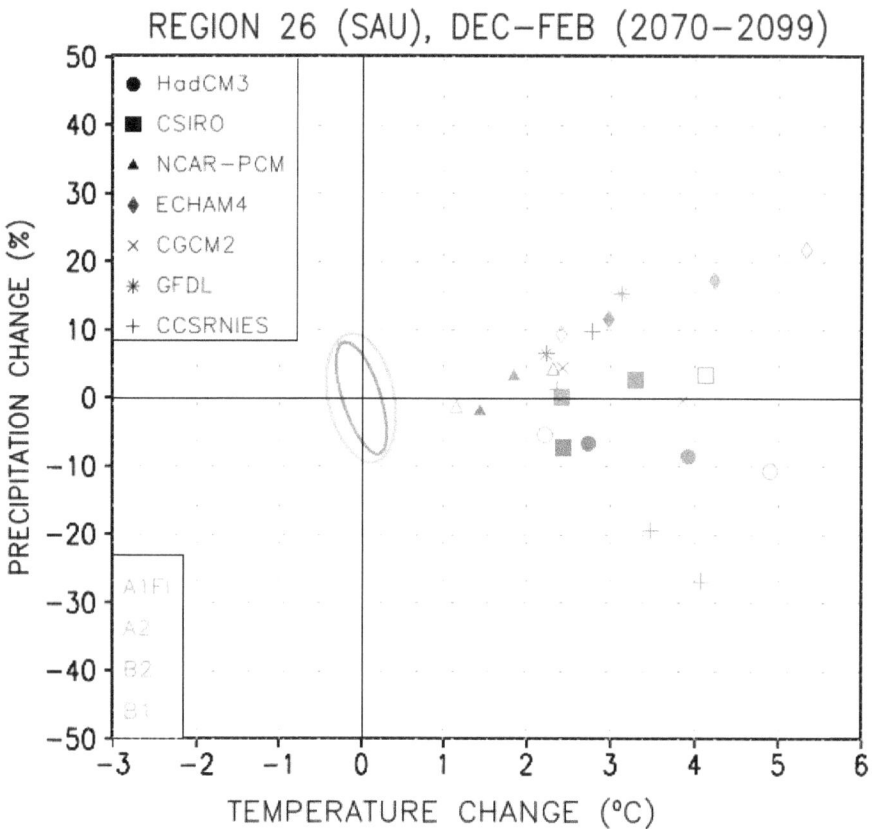

Figure 2.3. Scatter plot showing multiple model results (keyed to different symbols in the upper left-hand corner of the plot) for temperature and precipitation change for Southern Australia based on model solutions forced by different SRES emissions scenarios (different colors keyed to scenarios in lower left corner of plot), related to model estimates of variability (colored ovals). Such scatter plots are frequently developed to display seasonal information (Ruosteenoja *et al.*, 2003).

projections to current levels of climate variability that stakeholders and managers already have experienced in practice (Figure 2.3) (Ruostennoja *et al.*, 2003). Key issues in preparing the outlooks are rapidly establishing a regionally-based expert judgment process and providing some common information and assumptions about future global-scale changes on which the regional outlook groups could base their assessment. As a fallback, the NCA could simply make the data sets available to regional and sectoral teams, but prior experience in the NCA 2000 and 2009 indicates that this will reduce use of the scenarios.

2.4.2.3 Integrated sets of quantitative scenarios of climatic, environmental, and socioeconomic conditions
The standard information base for development and application of scenarios in climate research by intermediate users is quantitative data sets of model outputs produced by one set of models and provided to others as inputs. Approaches for these

inter-model and research community transfers have been refined with experience. Researchers from the integrated assessment, climate, and impacts research communities have established a new process for coordinating the handoff of scenario information called the "parallel process" to improve cross-community interactions (Moss *et al.*, 2010). This new process is likely to affect the options for producing the scenarios for the NCA. The process replaces a sequential approach in which detailed socioeconomic narratives and scenarios were prepared first to develop projections of emissions, which were then provided to climate models to produce climate scenarios, which were eventually provided for impacts, vulnerability, and adaptation research. The sequential process took many years to complete, resulting in inconsistencies, and the socioeconomic scenarios that started the process were usually focused primarily on energy supply or emissions projections. The parallel process reorganizes these inter-community transfers by starting from radiative forcing and developing

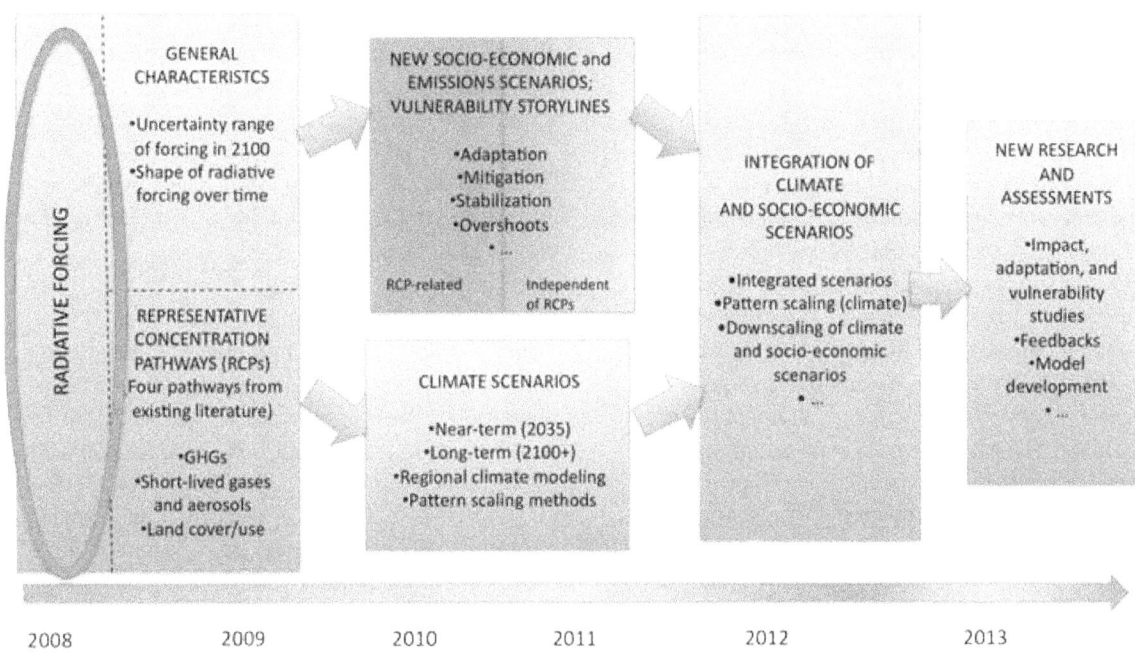

Figure 2.4. "Parallel process" for exchanging data and information across different research communities involved in climate change research and assessment (Moss *et al.*, 2010). General characteristics of radiative forcing (concentrations of greenhouse gases and other forcing agents over time) depicted in "Representative Concentration Pathways" (RCPs) are used in climate model experiments (organized under the auspices of CMIP5) and efforts to develop new socioeconomic scenarios (organized by a new Integrated Assessment Modeling Consortium, the IAMC, and by researchers who research impacts, adaptation, and vulnerability). The new process provides more time to develop socioeconomic scenarios that are conceptualized and developed to address questions related to both adaptation and mitigation. The new process is based on the observation that many different socioeconomic development pathways can be associated with any given radiative forcing trajectory. The new process has created many research issues and needs.

detailed socioeconomic scenarios and climate scenarios at the same time (Figure 2.4). This is enabling development of socioeconomic scenarios that address key uncertainties in factors that affect impacts, adaptation, and vulnerability as well as those that influence emissions, as has historically been the case.

Ongoing efforts are focusing on climate model experiments using the RCPs (coordinated as part of the Phase 5 Climate Model Intercomparison Project (CMIP5) of the WCRP) and development of socioeconomic scenarios (NRC, 2010). New scenario approaches for use by the IPCC were the subject of a November 2010 workshop "Socioeconomic Scenarios for Climate Change Impact and Response Assessments (WoSES).[2]" Within the context of the parallel process, a great deal of effort has been put into ensuring that emissions and land-use data associated with the

"Representative Concentration Pathways" (RCPs) were prepared and presented in a form readily useable by climate modeling groups. A so-called "handshake document" describes data provided by integrated assessment modeling teams for use in climate modeling (van Vuuren, 2008). Steps are needed to ensure that climate and socioeconomic scenarios are readily accessible to the impacts, adaptation, and vulnerability research community— an issue of great importance to the NCA in both the immediate context of NCA 2013 and the sustained distributed Assessment process.

2.4.2.4 Quantitative climate scenarios and downscaling

This section of the white paper focuses on potential needs and sources of data from different types of climate models and downscaling methods. Additional information on sources of data for environmental and socioeconomic scenarios will also need to be developed. The section draws heavily on an inventory of approaches to climate

---

[2] http://www.ipcc-wg3.de/meetings/expert-meetings-and-workshops/WoSES)

modeling and downscaling prepared for the Piloting Utility Modeling Applications (PUMA) for climate change workshop.

There is an increasingly sophisticated array of tools for developing regional climate change information for different uses, including dynamical downscaling using global and regional climate models, statistical downscaling, and historical climatologies. The data needs vary from use to use, as does the suitability of the techniques for discrete applications. A number of inventories of variables from climate model experiments needed for different types of impacts models have been prepared and are available. Moss and Marengo (2007) provides a list of variables requested by the impacts, adaptation, and vulnerability research community, and PCMDI (2011) provides the full list associated with the CMIP5 archive. This draft of the white paper does not catalog or prioritize across these needs because it is assumed that these needs go beyond developing contextual information and are related to modeling and other forms of quantitative analysis.

It is crucial to establish realistic objectives regarding provision and use of quantitative scenario information, especially in the context of NCA 2013. It may be the case that most of the quantitative information that is developed will be more useful to distributed assessments in regions and sectors, and thus to a future snapshot of this activity in the NCA 2017 Assessment.

A white paper inventorying the status and availability of data from a number of current climate modeling and downscaling efforts has been prepared for the PUMA project (Sharp, 2010). One of the major objectives of PUMA is to identify state-of-the-art climate modeling tools and techniques for use by a select group of Water Utility Climate Alliance (WUCA) members committed to being technically prepared to conduct climate impacts assessments for their systems. These members have both water supply and storm water management interests. The PUMA workshop products include a report from the meeting held in San Francisco December 1–3, 2010. The paper reviews the

Modeling Project				
	CMIP3	CMIP5	NARCCAP	RegCPDN
Approximate Resolution (degrees unless noted)	Atmos: 1.1 x 1.1 - 4.0 x 5 .0 Ocean: 0.2 x 0.3 - 4.0 x 5.0	native model resolution; details TBD	50 KM	25 x 25 KM (atmos only)
Output Timestep(s) Frequency	3 hrly; mon/daily mean; extreme	3/6 hrly; mon/daily/ annual mean	3 hrly; daily	daily; monthly; means; count[d]
Domain	global	global	North America	Western US
# Models	23	TBD	Regional = 6; Global = 4 (not incl. NCEP); 20 combo's planned	(1) Regional/ Global pairing - HadRM3P/ Had AM3P
# Output Params	118[a]	404[b]	49	50
SRES/RCP Emissions Scenarios	(3) A2, A1B, B1	(4) RCP's 2.6, 4.5, 6, 8.5	(1) A2	(2) A1B, B1
Time Periods Covered	1850 - 2000; 2000 - 2100; 2000 -2300	850 - 2300[c]	1980 -2004; 1971 - 2000; 2041 - 2070	1959 - 2010; 2010 - 2100 planned
Notes	basis for IPCC AR4 (2007)	basis for IPCC AR5 (due late 2013)		
[a] "High Priority Output" only; only ocean and atmosphere available [b] "Priority 1" output only; ocean, land, and atmosphere available [c] Range dependent on exactly which Tier 1 and Tier 2 experiments are selected [d] For example, Number of days with Tmax > 30°C				

Table 2.1. Summary of recent and ongoing model projects prepared for Piloting Utility Modeling Applications for climate change workshop (Courtesy of PUMA http://www.wucaonline.org/html/actions_puma.html).

Project							
	USBR/SCU	**Climate Wizard**	**NECIA**	**UWisc**	**USGS Cascade**	**CRU**	**UW CIG**
Resolution (degrees unless noted)	1/8, 0.5	1/8, 4 KM, 50 KM	city to regional (1/8)	10 mins	12 KM	10 mins, 0.5	1/16; 12 KM 36 KM
Output Timestep	Monthly	Monthly, Seasonal, Yearly	Daily, Monthly, Yearly	Monthly	Daily	Monthly	3/6 hrly, daily, monthly
Period(s)	1950 - 2099	1951 - 2006; 2050s; 2080s	1961 - 2099	1961 - 1990; 2041 - 2060; 2081 - 2100	1950 - 2099	1901 - 2002; 1961 - 1990; 1901 - 2100	1915 - 2006; 1950 - 2100; 3x100
Method/Algorithm	bias correct/ interp. (spatial)	Various (USBR/ SCU,CRU)	bias correct/ interp.; regress	bias correct/ change factor	construct-ed analogs	interp. change patterns, etc.	BCSD (Hy-brid) Delta; WRF model
Domain	US (1/8); Global (0.5)	US (1/8, 4 KM); Global (50 KM)	NE USA	Global	USA + Columbia R. (Canada)	Global	Western US; PNW
Emissions Scenarios	(3) A2, A1B, B1	(3) A2, A1B, B1	(2) A1F1, B1	(3) A1B, B1, A2	(2) A2, B1	(4) A2, B2, B1, A1F1	(2) A1B, B1
Params	precip, surface air temp	avg. air temp, precip	min/max/ avg temp; precip; extremes	avg. air temp and precip	precip, min/max temp	precip, wet days, temp, wind, etc.	temp, precip, winds, soil moist, etc.
Data Source(s)	CMIP3	USBR/SCU, PRISM, CRU, CMIP3	CMIP3 - GFDL, HadCM3, PCM	CMIP3	CMIP3 - PCM, GFDL	5 IPCC TAR models	CMIP3 (10 best)
Notes	48 or 112 scenarios 16 models					includes 20 change scenarios at 0.5	

Table 2.2. Selected downscaling projects from the Piloting Utility Modeling Applications for climate change workshop (http://www.wucaonline.org/html/actions_puma.html).

status of recent and current global climate model intercomparisons, including CMIP3 and CMIP5, as well as the North American Climate Change Assessment Program (NARCCAP) and the Regional Climate Prediction Dot Net project. It also reviews the status of selected downscaling efforts. Tables 2.1 and 2.2 summarize results of these activities.

In addition to quantitative climate scenarios and downscaling, quantitative scenarios of key environmental conditions such as land use and

sea-level rise will also be needed. Projection of environmental conditions sensitive to climate variables are developed using climate scenarios and data, and also serve as inputs to a wide range of quantitative and qualitative research and assessments that evaluate implications for human activities and infrastructure. For example, air quality is affected by anthropogenic emissions of a variety of pollutants and atmospheric processes sensitive to temperature and other conditions, and a variety of models use climate scenarios as inputs to model

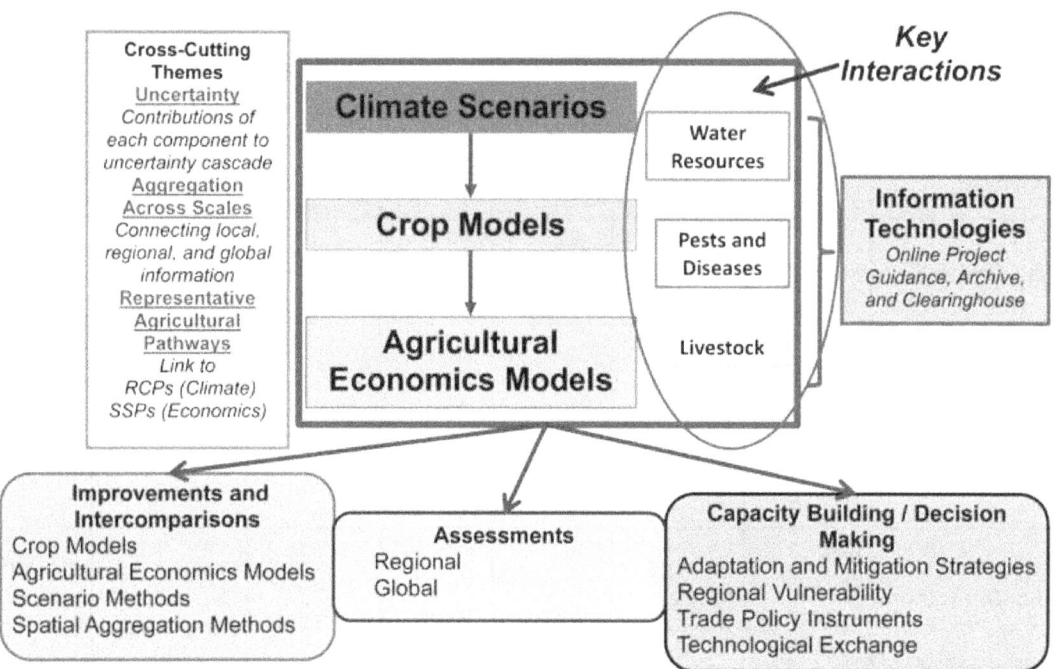

Figure 2.5. Conceptual framework of the AgMIP project depicting flow of scenario information from climate change, to environmental conditions and outputs, to models of human activity and impact (Courtesy of Alex Ruane and Cynthia Rosenzweig).

chemistry and circulation processes that affect air quality and incidence of air pollution events. Models of vectors of a range of diseases use climate scenarios as inputs and provide information used in assessment of human health impacts. Sea-level rise projections depend on a range of climate-sensitive processes and are used as inputs into studies of coastal erosion and flooding. Models of changes in outputs of different crops require climate scenarios and provide inputs to agricultural trade models that produce information on the potential impacts of climate change on agricultural prices and food security.

Quantitative environmental scenarios that examine the implications of climate change for human activities and well-being are becoming more sophisticated. Efforts at inter-comparing different realizations of models within related classes used to produce this information are still in the early stages, however. A current example of intercomparison activities is the Agricultural Model Intercomparison and Improvement Project (AgMIP). This activity is a distributed simulation exercise that will compare results across models for both historical and projected future climate change conditions with participation of multiple crop and world agricultural trade modeling groups. AgMIP will provide a multi-scale impact assessment using current

methods for climate and agricultural scenario generation. Scenarios and modeling protocols will be distributed on the Web, and multi-model results will be collated across crops and regions. Intercomparison of other types of environmental models and scenarios is needed in the longer term and may provide useful inputs to distributed Assessment activities (Figure 2.5).

2.4.2.5 Scenario tools to support participatory processes

Most global- or national-scale assessments are expert-driven, and as a consequence, scenarios developed to support these assessments have primarily been quantitative and used to coordinate different areas of modeling and evaluation by providing shared input assumptions. As mentioned above, however, there are significant benefits for end users participating in scenario development. There is growing experience, with participatory assessment approaches in which a mix of stakeholders and experts engage in a shared assessment process (NRC, 2009; Salter *et al.*, 2009). A participatory process is a purposefully designed set of activities structured around *framing* (including clarifying objectives and identifying participants), a set of *participatory activities* that can include workshops and engagement of participants through other means such as social media or technology such as decision theaters, and a set of *outcomes* that could be a decision, a community plan, a report, films or audios, or other forms of knowledge sharing or exchange.

Participatory processes for complex planning and decision making have been developed over several decades in response to shortcomings of purely expert-based decision support (Arnstein, 1969; Fischer, 1993). These processes enhance understanding and build community capacity for making informed decisions that can integrate scientific research and local knowledge. Participatory processes can enable participants to clarify different perspectives about potential impacts and response options and to build acceptance and ownership of agreed actions. But they also have costs, including requiring additional time and resources, and being more difficult to control with respect to focus or outcome. Research that inventories and evaluates approaches to development and use of scenarios in participatory processes for the context of the NCA is currently being prepared Wiek *et al.* (2010). This section of the white paper draws on this draft evaluation and briefly introduces several options and ideas for next steps.

The primary purpose of participatory processes has been the exchange or production of *knowledge* across different groups of experts and stakeholders (Wiek *et al.*, 2006). Following early categorizations, participation can range from *information* (communicating from experts to stakeholders) and *consultation* (eliciting from stakeholders to experts), to *collaboration* (mutual interaction, co-production). Standardized forms of engagement that correspond to these three categories are, for instance, *expert hearings/input* (information), stakeholder *focus groups* (consultation), and *workshops* (collaboration).

Participatory scenario studies on climate change develop or use the full spectrum of scenarios, from socioeconomic drivers and emissions to impacts and responses (Wiek *et al.*, 2010). Shaw *et al.* (2009), Loibl and Walz (2010), and Bryan *et al.* (2011) are three illustrative examples of how participatory scenario studies engage scientists and stakeholders in the development or use of climate change scenarios to anticipate local climate-change impacts and explore response options. Climate and socioeconomic scenarios are developed and used in these processes in a variety of ways that enable

participants to evaluate how local decision options that may be affected by changes in climate (e.g., related to community economic development, infrastructure, land use, investment in renewable energy technologies) perform under a range of potential future climate (and socioeconomic) conditions. In this sense, the ultimate purpose of many participatory scenario exercises is to help decision makers broaden the range of policies under consideration and to help choose among these policies.

However, the specificity needed in the participatory scenarios to develop this wider range of policy choices for decision makers might not be achievable across all regions and sectors for the 2013 NCA. That is, at higher levels of aggregation, there are likely too many variables and competing interests to effectively evaluate context-specific policy options. To address this limitation, the 2013 NCA might consider supporting several case studies that highlight this potential use of participatory scenarios, and future NCA efforts could expand upon these examples while providing the necessary information for carrying participatory scenario exercises with decision makers.

With advances in computer and communications technology, a new type of participation has emerged in scenario processes, namely participatory tools—first and foremost, interactive and immersive visualization tools (Salter *et al.*, 2009). These consist of a range of visual and spatial media derived from modeling, data, scenarios, and descriptive narratives used to contextualize climate change information in two and three dimensions at the local or regional level (Sheppard *et al.*, 2011). They go beyond conventional text and scientific charts, using specialized three-dimensional modeling software or widely available virtual globe platforms (e.g., *Google Earth*). Such visual tools have been shown to increase cognition (Winn, 1997), and improve the salience of information to users and awareness of experiential or qualitative factors (Nicholson-Cole, 2005; Sheppard, 2005). These characteristics address the need by decision makers to assess community acceptance and feasibility of otherwise sound decisions on adaptation and policy (Burch *et al.*, 2010).

Among the more specialized participatory tools which apply to climate change scenarios (with more or less sophisticated visual components) are (1) simulation tools that allow stakeholders to build capacity in systems thinking related to climate change drivers, impacts, and responses (e.g., Robinson, 2008) and (2) gaming tools that allow stakeholders to engage with climate change scenarios in entertaining and competitive settings Vervoort et al., 2010). Participatory tools can be integrated in participatory processes or can stand alone, for instance, as Web-based or kiosk applications that engage individuals (but do not enable direct person-to-person interaction). Advantages of participatory tools compared to participatory processes are standardized presentation of information, accessibility to potentially large numbers of users, instant feedback, and low- or no-cost usage. The downsides are the relatively high development costs, as well as the lack of in-depth exploration, deliberation, and adaptability to stakeholder interests. When tools such as visualizations are embedded in participatory processes, as in the Local Climate Change Visioning process, which integrates various types of modeling across scales within "visual narratives" (Pond et al., 2010), deeper levels of engagement and high credibility can be achieved with both non-expert and expert users.

Temporary and permanent facilities have been established to engage stakeholders in the development and use of climate change scenarios. Museum exhibitions, such as the recent exhibition on "Rising Currents" in the Museum of Modern Art in New York, provide temporary opportunities for stakeholders to explore climate impacts and response options to climate change. Compared to participatory tools, museum settings have the advantage that they allow for alternative forms of information presentation, such as large installations, dioramas, multi-media, experiential settings, etc. They also provide the flexibility to combine different forms of participatory tools and processes. "Decision theaters" have also been designed to support participatory scenario development and decision making related to climate change, on an ongoing basis. Using participatory tools and processes, in particular visualizations, decision theaters are both physical spaces in which the participatory process occurs and virtual spaces for decision support and evaluation research. More than a visualization production studio, decision theaters offer the advantage of a research laboratory (control, documentation, etc.), a resource for multiple end users to access standardized scenario data sets, and a hub for training in best practices (Sheppard, 2005); but there are also disadvantages

such as limited geographical accessibility, high maintenance cost, and required technical expertise. Permanent decision theaters are in operation or under construction at locations including Arizona State University, the University of British Columbia (Canada), University of East Anglia (UK), Linköping University (Sweden), and Huazhong University (China). An international research network among the decision theaters has been initiated.

This is a good foundation for expansion of the participatory use of scenarios in climate assessment. NCA 2013 could start by identifying and evaluating initial applications of scenarios in participatory processes with a shared, structured framework that allows a comparison of framing, participatory activities, and outcomes; for example with respect to success in engaging stakeholders from different regions and developing or using different scenarios to build and communicate knowledge. In the longer term, there are a number of tools and processes that the NCA could advance; for example,

- A handbook that offers information on a range of participatory approaches for constructing and using climate change scenarios (with empirical illustrations and case studies as templates for "good practice") and that describes their key features, strengths, and weaknesses (comparison and evaluation);

- A database with empirical participatory scenario case studies to provide a growing knowledge base and model projects for experts and stakeholder groups;

- A Web site that provides key information on participatory scenario approaches (short version of the handbook) and provides interactive exchange and research coordination (forum, blog, updating of the database, etc.); and

- Coaching and training workshops to build capacity in applying participatory scenario approaches.

2.5 Final Thoughts: Preparing for NCA 2013 Sustained Assessment Process

The NCADAC will need to address a number of crucial questions quickly if scenarios and related tools, methods, and resources for users and participants are to be developed in time for the 2013 Report.

How will scenarios be used by different sets of users in the NCA? The Assessment is more likely to succeed in its objectives if there is a strategy for preparing and applying scenarios. This strategy needs to be guided by the structure of the NCA 2013 report and plans for the long-term process. The structure of the report needs to inform decisions regarding time frames, spatial scale, uncertainties to be addressed in the scenarios, and technical guidelines for their use by the sectoral and regional assessment teams. The strategy needs to specify what products are required, who will use them, and how, so that materials can be prepared in a timely and appropriate fashion.

Support for implementation of the scenario strategy and monitoring progress throughout the process will also increase the likelihood that objectives are met. This could include providing facilitation to regional and sectoral teams.

The NCADAC should consider establishing a scenario working group composed of members of the research community and users to develop a detailed strategy for its approval. This working group could also provide support for users and monitor implementation. In addition, consideration should also be given to documenting and evaluating the entire scenario process for improvement of future assessments. The working group could assist in designing an evaluation plan.

What scenario products to support the Assessment should be prioritized? The four sets of products identified in the white paper constitute a good initial set of options for the Assessment. These include (1) socioeconomic narratives; (2) climate "outlooks"; (3) quantitative scenarios of emissions, climate, environmental conditions; and (4) tools and processes to support use of scenarios in distributed assessments. These materials and resources would provide both a degree of coordination across the Assessment and facilitate communication and stakeholder engagement to develop decision-support-oriented scenarios for the regions and

sectors. Many detailed questions remain about the specific attributes of these products, including

- What are the desired attributes of *climate* information that matter most? What timescales should be addressed? What variables are most needed? What are the most accessible and appropriate sources for these scenarios?

- What are the characteristics of needed *socioeconomic* and/or qualitative scenarios (e.g., future socioeconomic development pathways including such things as population projections, gross domestic product, land use, energy system evolution)? What are the sources of such information? How can scenarios be co-produced with local and regional expertise? What aspects of the future should be highlighted? What are the most relevant time frames for analysis?

- What *environmental scenarios* are needed (e.g., sea-level rise, air quality, and water quality and availability)? What are the sources for this information? How can consistency of information be ensured? How can the information be conveyed to users?

- How should uncertainty be represented in the scenarios (qualitative and quantitative options) in a manner that is transparent and useful? Can trends that are already inevitable be clearly distinguished by users from those that remain uncertain?

- What degree of consistency is needed across scenario components, and how can this level of consistency be achieved?

- What technical guidelines or information can be provided to facilitate development and use of regionally- or sectorally-oriented scenarios by teams in sectors and regions, and how can regional and sectoral activities be structured to identify priorities for tools, enhanced processes, and resources (e.g., participatory tools described above) for the long term?

- What data and information management systems need to be developed, for example to enable various users to access materials that are prepared?

Given the limited time available for preparing the NCA 2013 report to meet the requirements of the Global Change Research Act, the NCADAC will need to find an appropriate level of ambition for the Assessment strategy; one that balances wanting to provide a range of resources for participants over the short and long terms with the reality that time and resources are limited. The science of scenarios has advanced, and new tools and processes that facilitate application of scientific insights in deliberative and decision-making processes have advanced. As evidenced in this workshop, there is great enthusiasm among the research community and end users. An option for a minimalist strategy is to select from existing resources a limited set of scenarios on future climate, environmental, and socioeconomic conditions at a regional scale to facilitate coordination; to provide technical guidelines on how to relate existing research and other sets of scenarios used in the literature to the ones identified for the NCA; and to include an evaluation of selected completed or ongoing participatory scenario activities in the report. Additional options with a higher level of ambition include preparation of climate change outlooks and development of a process in which regional and sectoral assessment teams prepare scenarios that are embedded in the broad regional scenarios. Implementation of even some of these options will constitute an advance over past practice and contribute to preparation of the 2013 report and development of resources to support an ongoing distributed Assessment process—the key objectives of the NCA.

Chapter 3:
Summary of Workshop
Presentations and Discussions

Nathan Engle, Pacific Northwest National Laboratory and
Sheila O'Brien, U.S. Global Change Research Program

3.1 Introduction

This chapter of the workshop report summarizes the presentations and discussions that occurred at the workshop "Scenarios for Assessing Our Climate Future: Issues and Methodological Perspectives for the U.S. National Climate Assessment" on December 6–8, 2010, sponsored by the U.S. Department of Energy. The goal of the workshop was to explore the use of scenarios in the context of the National Climate Assessment (NCA). Approximately 65 people participated in the workshop, including researchers with expertise in climate and Earth systems; socioeconomics and integrated assessment; impacts, adaptation and vulnerability; regions and sectors; and participatory processes. Program managers from a number of Federal agencies that are part of the U.S. Global Change Research Program (USGCRP) participated in this event.

The program was developed with input from representatives of the science community and chaired by Richard Moss of Pacific Northwest National Laboratory's Joint Global Change Research Institute and Linda Mearns from the National Center for Atmospheric Research. A white paper was written to help calibrate thinking, frame key issues for the workshop, and lay the foundation for some of the significant elements of the NCA effort. The white paper was revised after the workshop to reflect participant comments and suggestions (see Chapter 2).

Significant workshop discussion focused on definitions and purposes of scenarios, and how to move toward a common understanding across a broad range of potential uses of scenarios in the context of assessing climate change. Presentations and discussions at the workshop served as an opportunity for participants to raise and discuss ideas concerning the use of scenarios in the NCA. Participants were given explicit instructions that consensus advice was not being sought by the workshop organizers or NCA staff.

What follows is a summary of workshop presentations, breakout sessions, and discussions. See the accompanying appendices for the agenda and lists of supporting committee members and meeting participants.

3.1.1 Welcome and Opening Remarks

Workshop chairs Richard Moss and Linda Mearns opened the meeting with a brief overview presentation of the purpose and agenda. The presentation helped to reiterate that the workshop was intended to identify options for consideration by the federal advisory committee for the NCA, particularly with respect to the types of scenarios and related products that would be useful (and used) in (1) the NCA 2013 report and (2) ongoing, distributed Assessment activities in places and sectors. Given the mix of participants from various research communities, federal agencies, and other relevant organizations, the presentation also emphasized the need to identify applied research efforts that would be useful for the ongoing, sustainable Assessment activities as well as options for preparing these products and maintaining relevant activities on timescales pertinent to both a near-term report and the ongoing process.

The remainder of the first morning and early afternoon was devoted to background and overview presentations that helped place this workshop into context for the various participants. The summaries of these presentations, and related discussions, are provided below.

3.1.2 Workshop Charge and Coordination Process

Robert Vallario, U.S. Department of Energy

The use of scenarios has played a major role in our understanding of global change at global and national scales; there is now an expansion of the use of these tools at regional and local scales. At the same time, there is an expansion in the timescales that are being considered, including a focus on "what if" projections that center on significantly shorter time frames than in the past—on the order of decades rather than centuries. In addition, the use of scenarios has expanded from inquiry-based climate research to inquiry, policy-making, planning, and decision-support research. Historically, mitigation and impact studies were conducted separately, while they are now often combined into mitigation, impacts, adaptation, and vulnerability studies that include consideration of multiple stresses and sectors. These increases in sophistication of scenarios make them more useful, but also more difficult to explain in some cases.

Scenarios are important for advancing science and updating the NCA with the latest scientific knowledge, capabilities, and methodologies. Scenarios, both quantitative and qualitative, will influence how we frame problems in ways that are meaningful to both researchers and decision makers. New topics and issues are being considered in this NCA. Scenarios can help focus attention and illuminate crosscutting issues. Scenarios are also useful tools for understanding interactions between social and physical systems.

3.2 National Climate Assessment Objectives, Structure, and Context

3.2.1 Overview of the National Climate Assessment: Needs and Objectives, Process and Organization

Kathy Jacobs, White House Office of Science and Technology Policy

The approach to this third NCA is different in several ways from previous assessment efforts; among the differences are (1) focusing on building a sustained and sustainable process, rather than on writing a single report; (2) using existing regional and sectoral networks to build capacity for doing assessments across the country; (3) developing permanent national indicators of change and consistent methodologies for assessment; (4) focusing on cross-regional and cross-sectoral assessments of risk and vulnerability; (5) deploying findings using Web-based platforms on an ongoing basis; and (6) building a foundation for decision support related to adaptation and mitigation. A strategic plan has been written and a large, standing federal advisory committee is in the process of being launched. Scenarios can play a significant role in this Assessment, including

- Helping to understand the consequences of actions (and inaction),

- Bringing global issues to local, regional and sectoral contexts—where people live,

- Serving as a tool for community engagement,

- Helping design an "internally consistent" set of assumptions to support decision making, and

- Building bridges between disciplines and between science and policy makers.

In parallel to the efforts of the NCA, the next iteration of IPCC assessments is also focusing on adaptation and decision support to a much greater degree than have previous assessments. The relationship between IPCC and the NCA requires some explicit focus, since these efforts are underway at the same time. One use of scenarios is to illustrate the wide range of societal framings that can occur. Thinking only about the emissions pathways and building nested components that are internally consistent provides a vast array of possibilities and a major research challenge.

An important contribution of scenarios is illustrating key vulnerabilities in the context of crosscutting topics. Scenarios are likely to be very helpful here, because there is little existing work in this area. Examples include a potential focus on processes (e.g., climate impacts on the nitrogen cycle). Other options within a "nested-matrix"[3] approach that is being proposed are focused studies in particular places, including issues like the water-climate-energy nexus or watershed-based analysis (e.g., the Columbia or Colorado Rivers or the Chesapeake Bay). Other options include the implications of climate change on rural communities or on indigenous peoples.

Existing investments in science across the agencies are expected to inform and support the national indicator concept. There is a need to focus on the architecture and the strategic part of this effort to analyze change at a broad scale and understand whether we are making any progress relative to the pace of change.

The Assessment is also being designed as a gap analysis to inform the science agenda and future science investments for USGCRP and other federal agencies.

There are multiple typologies of scenarios. Two types of scenario users, decision makers and intermediate users have been considered. These users have very different requirements. Decision makers often depend on scenarios that have been tailored to a specific decision context, to illustrate the possible consequences of alternative options. Intermediate users (e.g., analysts, scientists, contributors to assessments) generally use scenarios for strategic exploration of possible future conditions, and also as input to other scenario or modeling efforts. In past assessments, scenarios for intermediate users have been emphasized. For the NCA 2013 report, there is a need to determine the types of scenario products and activities necessary to support regional and sectoral analyses.

---

[3] A broad conceptual framework or matrix linked to smaller-scale illustrative examples (NRC, 2007).

Some intermediate users of scenarios depend on outputs from one scenario as inputs to another. It is not clear that this type of user should be a separate group— but each of these communities may need to clarify what is needed in the context of the NCA and discuss options for meeting these needs considering ongoing work.

3.2.2 Scenarios in the IPCC Assessment Report 5 (AR5)
Chris Field, Chair, Working Group II, Intergovernmental Panel on Climate Change

The scenarios process for the IPCC AR5 is an independent effort led by the scientific community. There is a catalytic role for IPCC with ongoing conversations to consider issues of mutual interests, consisting of informal agreements on a series of defined issues. Key inputs and processes associated with the AR5 scenarios have included an expert meeting in Noordwijkerhout, Netherlands, in September, 2007 (with Richard Moss and Ismail El Gizouli as co-chairs), an IPCC plenary that tasked WGII and WGIII to maintain coordination with the Integrated Assessment Modeling Consortium (IAMC), and several IAMC as well as joint IAMC and IPCC coordination sessions.

An important concept being considered for the IPCC AR5 is a new scenarios development process (i.e., a "parallel" process, described in a summary of Edward Parson's presentation, below) that is parallel

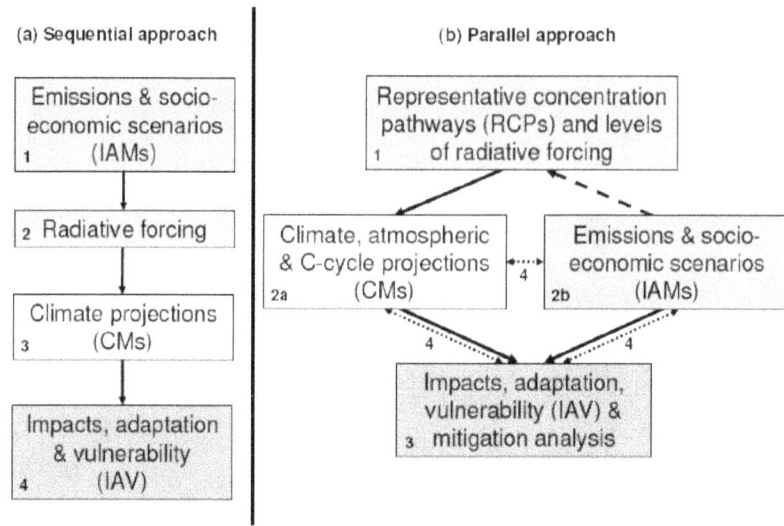

Figure 3.1. (a) Traditional sequential scenario development process, and (b) the parallel scenario development process. (Figure from Moss et al., 2008)

in nature, as opposed to the traditional sequential path (see Figure 3.1 for a depiction of the sequential versus parallel process, and Figure 3.2 for a more detailed view of the parallel process for the IPCC AR5).

The next step in the IPCC AR5 scenario development process is to create storylines—rich sector socioeconomic pictures that entrain a wide range of scenario producers. The storylines will help establish scenario "libraries", will pertain to market and non-market sectors (including ecosystem services, biodiversity, equity, and institutions), and will provide an avenue for feedback to emissions scenarios.

3.2.3 Post-Presentation Discussion

The point was raised that the national matrix of indicators concept could be very useful; however, there was uncertainty as to how the matrix might be informed by and relate to other federal efforts on climate change indicators. The general idea would be to make the most use of what is already being done by linking to and drawing from these ongoing initiatives. One of the main messages from the November workshop on ecological indicators for the NCA was to focus on the architecture and strategic elements of the design of the indicators

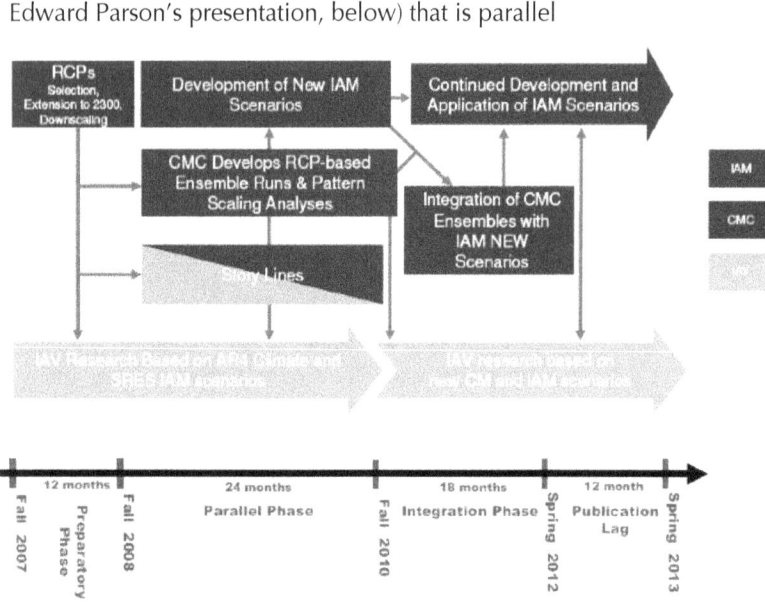

Figure 3.2. Detailed depiction of the parallel scenario process being pursued for the IPCC AR5. (Figure from Moss et al., 2008)

framework (e.g., determine which networks to draw upon, and identify good proxies for aggregation that can be used as indicators).

Participants also discussed the relationship between the NCA and the coordination of new research activities. Because of the need for quick turnaround for users, the NCA might not want to depend only on published literature; rather the NCA should be actively engaging in research and assessment as a continuing process.

Finally, the interaction between the NCA, the IPCC, and other assessment activities that use scenarios was identified as a potentially useful, but complicated relationship. Because of the wide range of framings possible in emissions pathways within these different contexts, there is a problem with trying to reach consistency between the activities. Nonetheless, the flexibility of the Representative Concentration Pathways (RCP) (Moss *et al.*, 2010) process might help to address this challenge.

3.3. Types and Uses of Scenarios

3.3.1 Overview of Scenarios in Climate Research and Assessment
Edward Hawkins, University of Michigan
There are a lot of definitional challenges related to the word "scenario." One useful definition is: a description of potential future conditions, developed to inform decision making under uncertainty. Scenarios need to be distinguished from other types of statements about future conditions, such as predictions, forecasts, and projections.

Relative to elements of a definition, these things called "scenarios" tend to be

- Richer, more complex, multi-dimensional (but still schematic and not exhaustive);

- Multiple, produced in groups; and

- Longer-term, extending further into the future.

Different types of scenarios can be linked in a linear, causal chain (e.g., socioeconomic scenarios → emissions scenarios → radiative forcing scenarios → climate scenarios → impact, adaptation, and vulnerability studies).

Though simpler, this linear approach does not represent actual processes very well, which is why there is now a proposal to use a parallel process

that accounts for feedbacks between human and physical system scenario elements. This process, referred to as Representative Concentration Pathways (RCPs), starts with radiative forcing scenarios, and allows for multiple pathways to the same outcome. Getting the scale and detail right across multiple complex processes is very challenging. However, we do not have a good alternative.

3.3.2 Climate Scenarios and Information
Linda Mearns, National Center for Atmospheric Research
Scenarios are used to explore deep uncertainties we cannot quantify (e.g., incomplete knowledge of physical processes, model structure, and important feedbacks within the climate system) and catastrophic extreme events (e.g., collapse of the Greenland Ice Sheet). Internal variability in climate modeling at the regional scale comes from different sources of uncertainty over time. In the early years it is dominated by model uncertainty, but over time it becomes dominated by uncertainties in the emissions trajectory. Models do not produce equally credible views of the future, and the metrics that can be used to evaluate them should vary depending on the questions that need to be answered (e.g., regarding scale, geography, or seasonality).

The objectives of downscaling are to bridge the mismatch of spatial scale between that of global climate models and the resolution needed for impacts and adaptation assessments or to resolve high-resolution processes that are responsible for regional climate.

Simple downscaling is adding large-scale climate changes to higher resolution observations (the delta approach). Statistical downscaling involves statistically relating large-scale climate features (e.g., 500 millibar heights) to local climate (e.g., daily or monthly temperature at a point location). Dynamical downscaling is the application of a regional climate model using global climate model boundary conditions. For the purposes of the scenarios workshop, there is not much need to focus on comparing various downscaling approaches and determining their value, but rather on how the downscaling approaches relate to the scenarios process.

3.3.3 Anticipation of Decadal Prediction Experiments for Use in Scenarios

Lisa Goddard, International Research Institute for Climate and Society

Changes in patterns of sea surface temperature (SST) are major drivers of climate variability. Understanding the mechanisms of decadal-scale evolution and persistence in SST patterns may hold promise for bridging the gap between seasonal to inter-annual predictions and long-term climate change projections, but much work remains to determine which aspects of the climate system are predictable at inter-annual to decadal timescales and whether current models and observing systems are adequate to realize such predictability. Even for climate change projections, there still are significant challenges to getting the patterns of regional trends right. Better understanding of decadal-scale variability and how that differs from anthropogenic trends is critical to improving predictive capacity. The International Research Institute for Climate and Society is investigating the possibility of "correcting" erroneous patterns of climate change projections due to model biases through statistical corrections and model weighting based on the realism of the tropical SST response to climate change. To help people make better use of models across timescales, we need to know intended timescale of information, what is being predicted, and what contributes to uncertainty at those scales. The future of prediction beyond seasonal predictions includes coordinated experiments for IPCC AR5 in working at the decadal scale and trying to understand the distinctions between climate variability and trends and how much of each is predictable.

What we have to work with now are CMIP3 climate change projections. These are adequate for temperature trends (if treated carefully) but precipitation is more problematic. We can also use characterizations of past variability to aid risk and vulnerability studies, particularly for variability on top of a slowly changing mean climate. In the longer term, we can make better use of observations, including statistical characterization of past climate variability and statistical interpretation of projected regional trends. Better dynamical models and higher resolution global models will also help researchers understand regional climate dynamics. However, as with any prediction systems, good models are only one element. Sustained observing systems, and appropriate methodology to incorporate those observations into models are equally important components if successful decadal predictions are to be realized.

3.3.4 Scenarios in Prior National Climate Assessments

Thomas Wilbanks, Oak Ridge National Laboratory

The first National Climate Assessment (NAST, 2001) provided:

- Historical records of past climate variability and change,

- Two climate change scenarios using large-scale British and Canadian general circulation models,

- Scenarios of changes in vegetation and biogeochemistry, based on the two climate change scenarios, and

- Projections of demographic and economic change plus narratives of possible technological and institutional change.

The second National Climate Assessment *Global Climate Change Impacts in the United States* (Karl *et al.*, 2009) used two categories of emissions scenarios—higher (either SRES A2 or A1FI) and lower (SRES B1)—to frame discussions of climate change vulnerabilities and impacts in regions and sectors.

In general, it appears that these scenarios were not readily used, especially in regional and sectoral assessments. Among the reasons that have been identified for the first National Climate Assessment were that the climate scenarios were delivered too late; that locally produced projections were more highly valued; and that the scenarios themselves were limited in scope, oriented toward averages rather than extremes, or were not actually answering user questions. For the socioeconomic scenarios, again there was a timing problem. Users were also not sure how to apply them for regional and sectoral assessments. These scenarios were mainly used in the synthesis report and in producing the report figures.

The scenarios for the 2009 report were used primarily for framing and communication. Higher and lower emissions scenarios, based on familiar SRES projections, illustrated the range of impacts, from moderate to more extreme. The intent was to communicate the importance of reducing emissions in order to limit impacts. The scenarios included regional downscaling and incorporated climate change/impact projections from other published sources.

Lessons learned from these experiences include the following:

- If scenarios are to be used in analysis and assessment, the approach must be understood and accepted by users at the outset;

- Scenarios need to provide information relevant to the vulnerability and impact questions being asked; and

- Scenarios must be delivered in understandable and usable forms early in the process of analysis and assessment.

Generally, this means that new scenarios cannot be developed for the production of NCA products that are on a tight time schedule. Even if provided in time, linkages between scenarios and impact literature will remain loose, because most of the published impact research will not have been based on those scenarios.

Scenarios combined with visualizations can be a powerful device for framing assessments of vulnerabilities and impacts. Often, such combined products are used to illustrate findings derived from other approaches, as they can help develop findings themselves. As a general rule, arraying findings from multiple scenarios is more informative (and believable) for risk perception and management than reporting results of a single scenario. Furthermore, scenarios incorporated in assessment reports seldom address questions about uncertainties.

Experience to date indicates that linkages between scenarios and impact assessment research are often indirect and sometimes entirely missing. Much of the published climate change impact research does not start by specifying particular quantitative scenarios; so trying to ensure consistency regarding scenarios would mean excluding a majority of the relevant impact research.

At best, it is sometimes possible to associate impact research results with climate change scenarios by connecting (1) changes (e.g., temperature and precipitation) posited in impact assessment research in order to explore sensitivities with (2) projections of such changes from scenarios. More often, the main value of scenarios has been contextual, to frame the questions being asked and to communicate results. One important issue that the third NCA will need to address is whether the assessment community can

set out to strengthen the linkages between scenarios and impact research in the long run.

3.3.5 Nested Scenarios: Approaches for Linking Different Scales of Analysis
Brian O'Neill, National Center for Atmospheric Research

Both processes and elements can be linked across scales; they are not independent and the choices depend on a scenario's purpose. Research-oriented scenarios often favor strong links across scales. This allows more comparability across studies in order to draw conclusions. Decision-support-oriented scenarios often favor weaker links, which allows more flexible regional scenario construction and promotes relevance and credibility at regional levels. Capacity-building scenarios favor process over product, with the specific content of scenarios being less important than the learning that takes place by participants.

Multi-scale scenarios are important because different processes (including climate drivers) act at different scales. Stakeholders and decisions at multiple scales are relevant; they require examination of cross-scale interactions and differentiating boundary conditions from decision variables.

Scenario processes, elements and outcomes can be linked across scales either strongly or weakly. A clear overall aim for the scenarios is crucial to deciding what the best degree of linkage is. A high degree of consistency across scales is not always necessary, or at least it is not generally the most important goal. The payoff from higher resolution modeling is not automatic and should be carefully considered in advance.

3.3.6 Post-Presentation Discussion

There was a discussion of the need in the third NCA for either a "central scenario" or a set of "bounding scenarios" without a central scenario. Key questions included (1) How many scenarios should there be, and (2) Does it need to be an even number? An even number of scenarios may be preferable, because when there is an odd number people tend to pick the one in the middle, missing the point of having a set of scenarios and explorations of uncertainties. For the NCA, there is likely value in setting several key assumptions for the more distributed process of creating scenarios for various purposes.

Another question for discussion was how the scenario process can help federal agencies with their adaptation activities, especially in light of the Interagency Climate Change Adaptation Task Force assignments to complete adaptation plans in the near term. An important component of the next NCA is capacity building, which is needed both inside and outside of the federal agencies.

Some participants also suggested that building new scenarios for the 2013 report is probably not possible, perhaps with the exception of a few targeted cases (likely based on user needs). It is more important to use scenarios that we already understand well and build from those for the NCA report in 2013. It is possible to imagine a longer-term process where both end and intermediate users help to develop the scenarios. This is a goal for the longer term and we need to design this process. "Light" versions of the Shared Socioeconomic Pathways (SSPs) products may be ready by the summer, which might be useful. Off-the-shelf stress scenarios could be considered as well. Using existing scenarios, grouped together to categorize existing results, might be useful for the short term.

Participants emphasized the need to engage people at a variety of scales. It could take until 2013 to develop the scenarios, and then it would take a few additional years to actually use the scenarios that have been developed. The NCA staff is contemplating a "scenarios cookbook" for near-term regional and sectoral workshops in 2011. If there is a standing review committee for real-time work, we may be able to shorten the typical peer-review and publication steps. This could be both top-down and bottom-up interaction with regional decision makers.

Because the NCA process cannot be all things to all people, the NCA federal advisory committee will need to prioritize a wish list and identify the biggest uncertainties around key issues that stakeholders are trying to address. For example, what degree of consistency is needed by various assessment teams? We cannot tell people what they should care about for their region or sector, but a handbook with some basic guidelines would be useful. Any handbook for users would have to be very clear about the different meanings of scenarios and their uses. There are clearly different needs for the 2013 report and for developing the long-term process for regional assessments.

Scenarios can focus on interactions between mitigation and adaptation, but this is challenging. Also, not all changes will occur gradually and there will need to be evaluation of state changes as well as changes in average conditions.

The point was raised regarding how climate-driven and non-climate discontinuities will play into future scenarios. There are extreme events both in the climate and in ecosystems, economics, politics, social systems, etc. The research community is interested in making progress on including multi-stress perspectives or a multi-stress matrix with guided sensitivity analysis. There is a need for including both climate and socioeconomic streams of stresses in developing scenarios for the NCA.

One idea is to try to connect SRES scenarios and the new RCP approach that is being taken by IPCC (Moss et al., 2010). Perhaps the NCA can create storylines or narratives about these connections. Mapping between socioeconomic scenarios and SRES is beginning to occur, but there are some scale issues with socioeconomic scenarios (there are fundamentally different processes occurring at global, national, and local scales), and questions remain (e.g., What are the local factors that in the aggregate have impacts on higher scales, and what information about the international or national systems is needed by local level decision makers?).

Discussion reflected the need to focus on the low-probability, high-impact events. Failure to consider some scenarios or impacts that are extremely unlikely, but consequential (e.g., extreme flood events and epidemics), will not serve decision makers well. Decision makers often want to be prepared for the worst-case scenarios. Most decision processes have a lot of conservatism built into them. As a research community, we move toward the middle ground because of aversion to being tagged as alarmist or speculative. The NCA will need a defensible process that allows it to talk about extremes that may have very low probability, but potentially catastrophic impacts. Having a plausible worst case is useful as long as one does not overly quantify things that are unknown. Another scenario approach worth considering, related to this issue of potentially catastrophic outcomes, is Granger Morgan's "inverse" version, which first identifies a dangerous threshold to cross, then examines what types of climate changes would lead to the crossing of that threshold.

Finally, the IPCC 2001 report on climate scenario development had sophisticated definitions of scenarios. Some workshop participants felt that the level of detail limited the utility of the scenarios, but that perhaps they should be revisited. There is a need to stay focused on the purpose of scenarios, as products and processes that serve different objectives within the NCA.

3.4 Regions, Sectors, and Crosscutting Issues—Assessment Priorities and Scenario Needs Presentations

3.4.1 Report from Sectors and Regions Workshop

Kate White, U.S. Army Corps of Engineers

The overarching messages from the regions and sectors workshop that are most pertinent to the scenarios workshop include the following:

- In many cases regional and sectoral assessments must be integrated because each alone is necessary but not sufficient. Sectors cross regions and regions integrate sectors. There is a need for some kind of continuum.

- The variability in crosscutting topics can lead to uncertainties that are best addressed by using scenarios to illuminate potential vulnerabilities to the range of outcomes.

- Scenarios need a compelling storyline that invites interest and establishes reasonable expectations.

- There is a need to emphasize what we know regarding climate impacts and vulnerabilities, rather than emphasizing uncertainties.

- Scenarios should focus on "What keeps you up at night?"—the outcomes that managers most want to avoid.

The presentation also identified several issues that are important, but have not yet been resolved, including the following examples:

- There are tensions between organizing the NCA around political versus biophysical regions. There are no boundaries that work for all types of analyses.

- There is also a tension between socioeconomic and physical sectors (e.g., coasts) and how to delineate integrated sectors (e.g., natural environment and biodiversity).

- How should topics, categories, and issues be selected for highlighting in the NCA? Criteria could include the rate of change; high sensitivity, vulnerability, or impact; emerging issues; areas definable as geographic regions; feasibility of projecting future states; scales for decision making.

- Regarding the delineation of regions, the group favored a hybrid approach that combines state-based regions with cross-regional analyses.

- The degree to which the NCA analyses should be based on integrated human and natural systems was discussed.

- The exact audiences for the NCA products and the process need to be defined in order to target the NCA effort appropriately.

- There were varied opinions about how to address topics including marine resources, and air quality, adaptation-mitigation links, and temporal and geographic scales.

Participants identified possible criteria for selecting cross-sectoral or cross-regional topics:

- Integration (new connections, new integration, strength of interactions)

- Timeliness (urgent, policy-relevant, real consequences)

- Relevance (broad interest "pull" spurs engagement)

- Capacity or readiness (actionable for users, direct research and development, short-term NCA 2013 report, long-term requiring foundation now)

- Represents new understanding of systems; contributes to set-up for future assessments

Again, there was a sentiment amongst participants at the regional and sectoral workshop that one key criterion for prioritization of topics in the NCA 2013 report could be "what keeps you up at night?" Asking stakeholders and scientists this question could help identify unanticipated and unintended consequences; abrupt change; thresholds and tipping points; compound events; cascading impacts, etc.

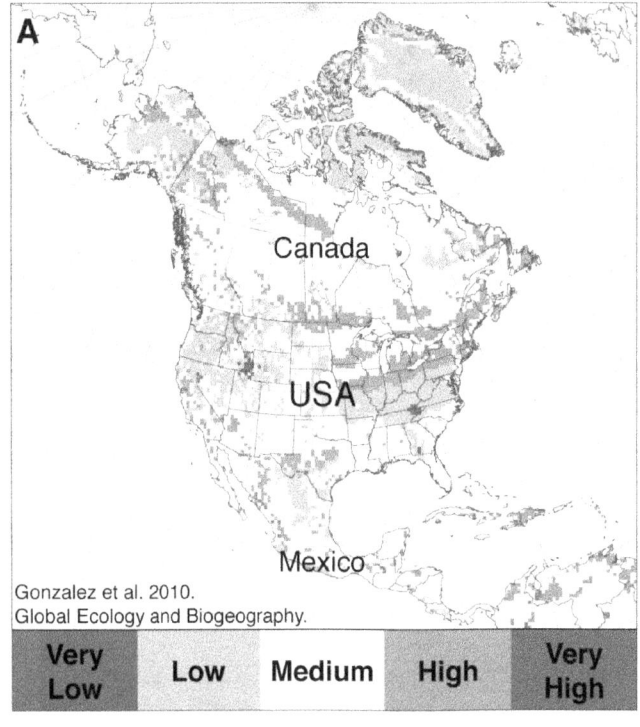

Gonzalez et al. 2010.
Global Ecology and Biogeography.

Very Low	Low	Medium	High	Very High
(0-0.05)	(0.05-0.20)	(0.20-0.80)	(0.80-0.95)	(0.95-1.00)

Climate Change Scenarios for Resource Managers
Lakes Michigan and Superior

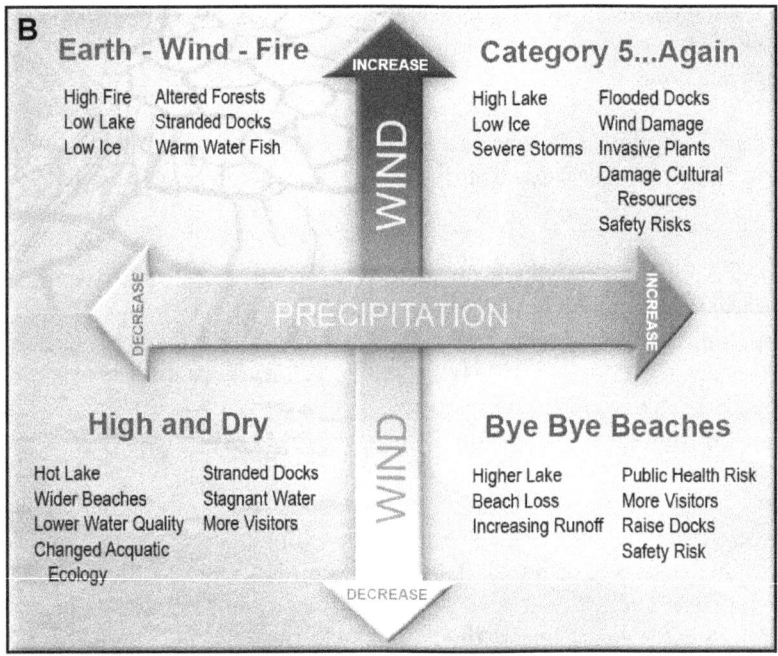

Figure 3.3. Examples of National Park Service use of scenarios for climate change adaptation. (A) Vulnerability of ecosystems to vegetation shifts based on 20th century observed climate and 21st century vegetation projected under nine combinations of emissions scenarios and climate model results (Gonzalez et al., 2010). (B) Management scenarios for park units on Lakes Michigan and Superior. Interdisciplinary teams developed four scenarios based on wind and precipitation - climate variables that are important for resource management but which exhibit large uncertainties. (Figure redrawn with permission from Patrick Gonzalez)

Finally, there was some concern among stakeholders that focusing on uncertainty in climate assessments diverts decision makers' attention from real challenges and knowledge about how they can proceed based on what is known today. However, different sectors are in different stages of readiness (e.g., water managers are ready to act based on what they already know).

3.4.2 Report on User Needs Identified in the Piloting Utility Modeling Applications Workshop
David Behar, San Francisco Public Utilities Commission
An example of a sectoral organization that is already tackling climate modeling and uncertainty issues is the Water Utilities Climate Alliance, a group of 10 large utilities that currently serves 43 million customers in the United States. They have been working together to identify research needs, evaluate next steps, and compare notes about use of models and scenarios. They are focused on identifying actionable science.

Some key conclusions follow:
* One size never has and never will fit all.

* Defining vulnerabilities and important management issues at the local decision-maker level is a good place to start defining information needs.

* Best past assessments have been done iteratively and sustainably and have exhaustively involved all players.

* Climate modeling tools are still elusive.

* The choice between CMIP3 and CMIP5 results is a bit fraught for users who find it difficult to know how to handle the decision of using well-vetted products versus the latest and greatest in model results.

* Beware the "Qualitative Paradox" (it is going to get drier, more intense, etc.; whereas quantifiable change is really what matters).

* Uncertainty is still the elephant in the room, but is more of a barrier for some decision makers than others.

* Sewer sheds (particularly for flood impacts) are often tougher to deal with than water supply or drought.

It is useful for utility managers to gain greater perspective on uncertainty and implications for action. Managers sometimes say they would just like to know the trend, but others focus on extremes. What are they really asking? In the case of drought for example, there is research indicating that drought is going to be more frequent. In the face of little predictive capacity, future climate could be portrayed as something similar to a historical worst-case event as a placeholder. It is valuable to use a longer period of record to illustrate a need to be ready for more frequent or longer duration droughts. This same logic applies to floods; better characterization of the climatology is something water managers need.

Finally, water managers know uncertainty and work with it every day on different timescales (e.g., daily, monthly, or annual cycles). However, their thought process is based on stationary historical records and demand trends. There is no good answer on how to package uncertainty in a way that all decision makers can understand.

3.4.3 Application of Climate Change Scenarios to Natural Resource Management
Patrick Gonzalez, National Park Service
Resource managers from the National Park Service (NPS) and other agencies are using climate change scenarios in two steps of the adaptation process: vulnerability analyses and scenario planning. The NPS analyzes vulnerability by combining 21st century projections of climate and ecological variables based on IPCC emissions scenarios with 20th century observations of climate and ecology. The NPS is analyzing the vulnerability of ecosystems to vegetation shifts and of several animal species to climate shifts.

The NPS uses spatial data from the vulnerability analyses in scenario planning, a process in which interdisciplinary groups of resource managers and scientists develop future management scenarios that might confront NPS staff in the future (Figure 3A). The groups examine pairs of climate variables that are important for resource management and exhibit large uncertainties. A pair of climate variables defines four possible future scenarios that are plausible, divergent, and challenging (Figure 3B). For each scenario, the groups develop adaptation measures that can respond to each scenario, producing a flexible set of options available for managers as conditions unfold. The NPS has conducted scenario-planning sessions in six landscapes and trained over 100 staff in the method. Scenario planning sessions are planned for more landscapes.

Data needs for management scenario planning include observed and projected climate and ecological data downscaled to an appropriate spatial resolution, measures of both central tendency and variability for each variable, probabilities of emissions scenarios, and quantified uncertainties of climate and ecological observations and projections.

3.4.4 Interaction of Sectors and Regions with Climate Scenarios
Commander Blake McBride, U.S. Navy
The Earth's climate is changing and these changes are most dramatic in the Arctic. The U.S. is an Arctic nation and the U.S. Navy must be prepared to respond to changes in this region. Scenarios are useful tools that can help the military determine how to respond to given stimuli. The key is defining these stimuli in a useful and appropriate way. The U.S. Department of Defense (DoD) uses scenarios to help articulate plans in the face of uncertainty to provide capability for the entire spectrum of potential contingencies. Fundamentally, scenarios are used to engage the imagination of decision makers and support policy changes. Scenarios help illuminate areas that are of most concern. Some aspects of climate change are easier for decision makers to understand; for example, sea-level rise is going to impact coastal bases in 50 years, and this is easily communicated. In contrast, the importance of ocean acidification might be more difficult for senior leadership to understand in the abstract. The DoD applies the same deliberate process for any challenge. The outcomes and insights that can be gleaned from using these scenarios can be used to properly focus research and development.

Related to the previous presentation on the National Park Service process, the DoD develops and approves scenarios regarding parks in the same regions independent from those of the NPS. This begs the question; is there a need or desire for a more integrated, cross-sectoral view among federal agencies and departments?

The Navy's strategic plan for the Arctic is based on avoiding conflict in the context of geo-political changes in the region, and wanting to work with others to show how climate change might play out in the Arctic. Also, transparency engenders trust, and working across agency boundaries could save money. Having partners with studies that they can share is valuable. The DoD cooperates with NOAA and the National Marine Fisheries Service and the National Marine Sanctuary Program, but the DoD has not traditionally been a big partner with the NPS (except for Southern California, where there are large bases contiguous to National Parks).

3.4.5 Post-Presentation Discussion
The panelists were asked how scenarios can work for such a broad set of regions and sectors planned for the NCA. As the DoD and NPS exemplify, scenarios can be used in a variety of contexts, and can contribute to a more integrated, cross-sectoral view of climate impacts, vulnerabilities, and decision making (e.g., defense and natural resource sectors in a single location). Having the agencies work together on scenarios will increase transparency and in turn engender more trust from stakeholders; save money by working more efficiently; lead to more integrated understanding of trans-boundary issues; and improve management decisions. Notable examples include NPS cooperation with NOAA on managing National Marine Sanctuaries, and DoD cooperation in Southern California with NPS, where large bases are contiguous with National Parks.

3.5 Breakout Discussions I: Possible Uses of and Needs for Scenarios in the National Climate Assessment

Three parallel breakout groups discussed needs for and uses of scenarios in the NCA. The following issues were considered:
1. Different desired attributes of climate information (contextual information, observations, and projections) for NCA 2013 report and ongoing sustainable process.
2. Needs for socioeconomic narratives and scenarios (qualitative descriptions of different socioeconomic development pathways)—at what scales and focused on what attributes of the future for the NCA 2013 report and ongoing sustainable process?
3. Needs and uses of environmental scenarios (e.g., sea-level rise, air quality, water quality and availability, and land use) for the NCA 2013 report and ongoing sustainable process.
4. Needs for data and information handling—how should scenarios be provided to both intermediate users and stakeholders for the NCA 2013 report and the ongoing sustainable process?

The breakout groups discussed the four needs outlined above for the remainder of the first day and for the first hour of the second day, and then presented their ideas to the larger workshop group. Highlights of group discussions and recommendations follow.

Group 1. Facilitated by David Behar, San Francisco Public Utilities Commission
This group synthesized priority information needs through a list:

1. Scale-appropriate (local, regional) socioeconomic and climate information together to address user-community needs, and good-practice guidance on how to downscale and use information, including
   - A data system of downscaled climate information and their inputs;
   - Evaluation of climate models, downscaling methods, and impact tools;
   - Focus on educating users as to the most useful application of climate data;
   - More explicit hand-off between intermodel comparison projects (CMIP to RMIP to IMIP, both ways); and
   - Design nested and scalable information.

2. Where possible and useful, explore probabilistic assessment of model inputs (likelihood of occurrence of one outcome versus another) to improve on expert elicitation, including
   - Estimation of probability that scenarios permit the quantification of uncertainty, which could be contentious;
   - Quantitative characterization of climate outlooks, but qualitative characterizations (e.g., narratives) are useful when not possible;
   - As an alternative to probabilities, identify strategies that are robust under a range of scenarios;
   - Where it is not possible, an explicit discussion would be useful; and
   - The context for climate information needs.

3. Decide whether policy should be included as a type of scenario; if so, incorporate policy and policy-relevant information, including

- Policy scenarios—CCSP Synthesis and Assessment Product 2.1A considered control scenarios (controls on emissions). The IPCC Fourth Assessment Report does not include control scenarios.

- Policies will lead to different pathways (energy policy, land use, etc.), impacting vulnerabilities.

- Connect scenarios to user concerns.

4. Illustrate threshold-type changes by communication about extremes and use of historical examples to discuss the increasing likelihood of threshold-type changes to various audiences.

5. Build in feedback about how to improve the process over time, including
   - Doing a better job with scale issues and separating focus versus form of scenarios (e.g., narratives and outlooks versus climate and socioeconomic scenarios);
   - Characterization of uncertainty that will allow users to more effectively assess vulnerability, make adaptation plans, etc.;
   - Better use of observations, which is an important part of making scenarios more useful; and
   - Identifying how to relate CMIP5 to previous work (e.g., CMIP3).

6. Conceptual framework—need a structural model of how different sectors intersect or interact and how to think about the connections, considering
   - Integration in general and multiple stressors in particular as an integrator of themes (i.e., climate is one driver, but there are many drivers and feedback mechanisms).
   - Focus on new sectors, cross-sectors, and multi-stress impacts that consider
     o Non-climate changes as important as climate changes;
     o Defining scenarios without climate and then deriving the outcome given climate;
     o Differences between outcomes and thresholds—need both;
     o Environmental scenarios that are consistent with climate scenarios;

- Understanding the integration between different NCA efforts (e.g., modeling, scenarios, and indicators);
- Focus on cross-sectoral nexus issues (e.g., water and energy);
- Strong use of observations (standard indicators and measures for modeling evaluation); and
- A national center for climate assessment that focuses on public engagement and cross-lab, cross-discipline, and cross-community efforts.

In addition to these priority information needs, there is interest in knowing why downscaling is of so much interest to a broad group of stakeholders. A second question relates to scale appropriateness and the integration of scale across different types of data (e.g., socioeconomic and climate). The NCA will need to make judgments about scale-appropriate information. Probability density functions can be very helpful, covering global and national policy, but tools short of the level of specificity provided by probability density functions may be worth considering. This leaves open the question of assigning probabilities. Perhaps characterizing the uncertainty from an end-user perspective would be useful. Many people do not understand the difference between confidence-level statements such as *very likely* and *virtually certain*; anything that would simply classify things that are definitely happening, or where models are unclear, would be very helpful to users. Overall, there is a need to simplify uncertainty.

Group 2. Facilitated by Anne Waple, National Oceanic and Atmospheric Administration

This group also synthesized priority information needs as a list:

1. Users and types of scenarios depend on context and decision makers (intermediate- versus end-user needs).
2. A portfolio of types of scenarios—short-term (S) versus long-term products (L)
   - Climate outlooks for 25-year and century scale (S),
   - Historical climate analogs (e.g., Dust Bowl) (S),
   - Transposed climate (e.g., moving states maps) (S),

- "What ifs" based on less certainty (e.g., quadrants) (S),
- Integrated scenarios: climate-impact-societal (L),
- Nested scenarios (L),
- Socioeconomic narratives (L),
- Quantitative scenarios (L), and
- Scenario-based decision-support tools (L).

3. Rapid assessment of scenario use in participatory processes (case studies)—again, short-term (S) versus long-term products (L)
   - Identify short-term deliverables as part of long-term strategy (S), and
   - Development of capability for use of scenarios (S, L).

4. Technical guidance products (e.g., which scenarios different users find useful)—again, short-term (S) versus long-term products (L)
   - Common requirements (e.g., a guidebook) for regional and sectoral groups to identify comparability and communication in synthesis (S),
   - Select priority "narrative" scenarios for analysis (S),
   - Limited number of case studies at smaller scales provides tests for longer-term processes (S, L), and
   - Central resource and technical guidance for downscaling and scenario development (L).

5. NCA scenarios should focus on a shorter time frame, 10 and 25 years, but maintain long-term predictions (e.g., infrastructure decisions are multiple decades)
   - Leverage existing efforts and networks
     - Existing SRES scenarios,
     - New IPCC scenarios might not be ready for 2013, as well as
     - How to relate CMIP5 to previous work (e.g., CMIP3)?

- Scenarios for 2013 NCA
  - o Parallel approaches (e.g., common climate narrative elements and model-based evaluation of narratives), and
  - o Focus on new sectors, cross-sectors, multi-stress impacts.

- Scenarios in the long-term Assessment process
  - o Parallel process with 2012/2013,
  - o Shared framework for classifying scenarios to enable local scenario and storyline development, and
  - o Dependent on communities-of-practice collaboration.

6. Additional important considerations
   - Scenarios are an important engagement tool;
   - Given the variety of scenario tools, characterization of uncertainty may be different in each type of scenario;

   - Integrated scenario development process can drive the research community forward and help stakeholders overcome uncertainty issues;

   - Visualization is an important component;

   - Multiple modes and levels of communication are important to reaching stakeholders;

   - Provide a range of possibilities, experiences, and information (i.e., to bracket the uncertainty);

   - Sensitive to surprises and game-changing events;

   - Provide tool box and clear guidance on the limitations of the products; and

   - Tools that provide actionable information to support decisions (e.g., raw climate information versus refined indices).

Scenarios can be used in combination with an understanding of the past to help bracket all possibilities in the future. It is important that whatever happens falls within those bracketed possibilities. Therefore, they must be sensitive to surprise; we must be willing to take on very unusual scenarios and consider unintended consequences and multiple stresses.

In the first round of iterative scenario processes, scenarios must stoke the imaginations of the stakeholders; let them guide the user to the questions. A good approach is an evolving conversation within a consistent framework for the largest scale that can be modified for specific needs. That said, sometimes stakeholders need specifically tailored pieces of climate information to make specific decisions.

There are, however, some potential pitfalls with this approach. For example, it will be important to avoid the ecological fallacy; if we create large scenarios at the national scale, we cannot apply that same logic at the local scale in all instances. Also, we still need well-developed toolboxes for downscaling. Thoughts for the NCA include coordinating existing work and recognizing that a number of agencies and businesses already have their own scenarios. This should be an iterative process, and NCA needs a leadership role with straw-man and zero-order discussions at the local level.

There is a need to stress more sophisticated interactions among socioeconomic variables that will describe possible futures. If we look only at existing SRES and related reports, in haste to be generalizable and robust, we will have put together very simple portraits of the future. Hopefully in the next iteration we have an opportunity to drill down and be more sophisticated and elegant in specifying alternative futures at regional and local scales within specific sectors. When we look at the future of society-environment-climate interactions, a number of changes will occur irrespective of climate change. It is important that in scenarios defined in the future that we include all families of environmental change and anticipate critical thresholds and rates of change as the environment evolves with climate.

In business schools, one of the prime teaching tools is the use of business case studies in which a set of boundary assumptions or conditions are put before students who are asked to build a scenario and response to a problem or challenge. It might be useful to take advantage of the tools used to develop these kinds of case studies to get started on scenario building.

3.6. Participatory Scenario Processes Presentations

3.6.1 Scenarios in Participatory Processes: Connecting Stakeholder Concerns to Global Change

Gregg Garfin, University of Arizona

We have learned important lessons from several stakeholder engagements in the context of Southwest water management: there are some tricks to the trade. For example, helping to create a community and getting to know other participants is critical to successful participatory processes. Also, there is a lot of interest among water managers to know what the plausible range of futures is, and they already have some mental maps of what is plausible that may not match with what science says. Some agencies still have low tolerance for uncertainty and prefer deterministic estimates. Thus, addressing uncertainty and presenting uncertainty information in ways that are familiar to stakeholders will address concerns about credibility.

Stakeholders also want help in overcoming "climate change planning paralysis." Approaches include mutual learning to collectively understand a system through conversation and interaction (which establishes a safe environment and allows for new voices and trust-building) and increased capacity to act. Stakeholders see structured scenario planning as an effective process for examining potential impacts and response strategies and to explore ways in which current management strategies can be used more frequently, extensively, or altered to meet future challenges. Significant limitations to participatory processes include the hard-to-overcome tensions in values and expectations as well as the difficulty to generate thinking "outside of the box."

3.6.2 Use of Scenarios in Metro New York City and New York State Assessments and Planning

Cynthia Rosenzweig, NASA Goddard Institute for Space Studies

Climate change planning in New York City and the surrounding region was strengthened in September 2006 when Mayor Michael Bloomberg created the Office of Long-Term Planning and Sustainability, with the goal of developing a comprehensive plan to create a greener, more sustainable city. Mitigating and adapting to climate change are central goals of the City's comprehensive sustainability plan, PlaNYC, which was released in April 2007.

In addition to its goals to reduce greenhouse gas emissions, PlaNYC includes adaptation as it recognizes the importance of doing both climate change mitigation and adaptation simultaneously to protect the citizens of New York City. One climate change adaptation goal of PlaNYC is the creation of an interagency task force to protect the City's vital infrastructure in the face of a changing climate. To meet that goal, Mayor Bloomberg convened the New York City Climate Change Adaptation Task Force in August 2008. The charge of the Task Force is to identify climate change risks and opportunities for the City's critical infrastructure and to develop coordinated adaptation strategies to address these risks. The Task Force consists of approximately 40 city, state, and federal agencies, regional public authorities, and private companies that operate, maintain, or regulate critical infrastructure in the region. The Task Force is developing climate change adaptation strategies to mitigate the risks posed by climate change to the City's critical infrastructure related to energy, transportation, water and waste, natural resources, and communications. A key outcome will be a comprehensive citywide plan to increase the City's climate resilience.

To support the Task Force, Mayor Bloomberg (in partnership with the Rockefeller Foundation) convened a group of climate change and impact scientists and legal, insurance, and risk management experts as the New York City Panel on Climate Change (NPCC), which was also launched in August 2008, to advise the City on climate change and adaptation.

3.6.3 Making Local Futures Tangible: Synthesizing, Downscaling, and Visualizing Climate Change Scenarios for Participatory Processes

Stephen Sheppard, University of British Columbia

Visualization techniques are emerging as important tools in scenario planning. In research studies using science-based visualizations of future scenarios to engage local stakeholders, the idea that communities have choices to make in both mitigation and adaptation is new to many participants (Sheppard et al., 2011).

More research is needed to assess effectiveness of these tools and processes with more user groups, but evaluation studies to date have shown a high level of engagement in developing and interpreting future scenarios, ready adoption and understanding of the scenario framework by the public, credibility of the resulting visualizations, and an increased awareness of local climate-change impacts and possible solutions. The visualizations in particular appear to provide a more motivating experience for stakeholders.

Conclusions for the NCA from this work are that holistic, credible, engaging stakeholder processes on climate change can be successful at local and regional levels. People appreciate simple scenario frameworks they can grasp, from which they can develop more specific local or sectoral variants. Visualization can rapidly increase awareness in both directions between scientists and end users (e.g., co-production of knowledge on appropriate adaptation and mitigation solutions); and visioning processes can be useful in developing salient scenarios and assessing socioeconomic and qualitative implications (e.g., cultural constraints on technical solutions). More research and case-study applications are needed on developing and evaluating hybrid modeling processes and integrative mapping for their effectiveness and transferability. Where downscaled information is not yet readily available for local users, iconic visualizations of climate change impacts, adaptation, and mitigation options in typical regional conditions can be effective for informing users.

3.6.4 Connecting Stakeholders to Global Change Futures: Gaming, Visualization, and Other Methods
James Buizer, Arizona State University
Knowledge systems must be perceived by stakeholders as credible, salient, and legitimate. One way of achieving this is through visualizing climate change futures. The Arizona State University *Decision Theater* is a good example of a tool to help decision makers in this visualization process, particularly to study water management decisions in central Arizona in the context of rapid population growth and urbanization, complex political and economic systems, variable desert climate, and the specter of global climate change.

Another example is WaterSIM, which is an integrated simulation model that uses exogenous uncertainties, policy levers, relationships, and other factors to simulate water consumption and availability in central Arizona from the present until 2030. WaterSIM has been tested with multiple audiences. Interestingly, it is least trusted by consultants (reasons may be either their high level of expertise leads to skepticism or that they have tools and business

practices that are in competition with this effort). There are a lot of people who profess to know how to translate science for decision contexts but there are not that many successful examples. Thus, there is a real need for boundary organizations and objects to address the differing needs and concerns that various stakeholder groups express.

3.6.5 Post-Presentation Discussion
There was a discussion about what exactly represents actionable information in participatory scenario processes. Agencies and organizations want something they can use right away. They want early wins, which do not necessarily require detailed and exact quantitative information. Climate risk factors can help.

In New York, there has been a fascinating connection of science with decision making. An important challenge has been how to decide upon which part of the probability distribution function to base planning decisions. For example, what level of risk is acceptable? Stakeholders in New York decided to use the "worst case" plus rapid ice melt, which highlighted the need to move pumps to deal with significant sea-level rise. In the New York area, protecting against the maximum of projections seemed to be where they are headed.

Some water management agencies in the Southwest have models and deal with climate issues on a regular basis; despite the demonstrated value of using ensembles, some refuse to use an ensemble as an option. They perceive that they will lose too much of the variability, and extremes are too important. They did not want to play games with statistical decomposition and then add variability back in (though we might be able to convince them of that when describing tree-ring data since they are very comfortable with this data source). To have a politically comfortable range, water managers in the Southwest added a very wet model that scientists had no confidence in. However, when the dynamical downscaling showed a change in sign from the statistical approach, they were very skeptical about the results of the entire effort. Clearly there is a need for translation and ongoing engagement between stakeholders and scientists so that the interface is very carefully managed and expectations are realistic. It was not unexpected that the two methods would lead to different conclusions, and in fact they had been warned that this could happen.

There was a question as to whether in using global emissions scenarios to drive regional scenarios in New York City, there was (1) any effort to reconcile a global picture with a local one in the context of variables, (2) any consideration of uncertainties (socioeconomic assumptions, regional considerations, etc.), and (3) any regional information that would be helpful in thinking about using a global context to frame local contexts. It was highlighted that only in the "soft links" were global SRES emissions linked to regional scenarios in this particular New York City exercise. In earlier work, the Environmental Protection Agency Star project with a New York City climate and health project actually explicitly evaluated linkages from SRES population and gross domestic product (GDP) projections to the modeling done for the region with population, GDP, and land-use projections. Regarding uncertainties in the socioeconomic assumptions, the New York City mayor worked with McKinsey Consultants to obtain the relevant socioeconomic data. Finally, linkages from global to national would be very helpful, but only "soft links" from the regional to global were considered.

In Vancouver, they used a combination of soft and hard links, mostly soft. They anchored scenarios with quantitative downscaled information and built on that approach for a core of credible scientific information. At the regional level, global-local equivalency was assumed for global emissions scenarios and local mitigation efforts, using more qualitative coherent storylines that matched across scales.

- During the discussion, there was a request that the panelists provide one suggestion each for what the NCA process should do in the next six to nine months to improve the quality and value of the 2013 report, and their responses are summarized as follows:

- We should have assessments carefully focus non-climate impacts on resources that we are managing and demonstrating in our assessment; we need a good baseline for such an effort, upon which we can then overlay climate impacts.

- There is insufficient time to do a proper needs assessment among stakeholders or an evaluation of different approaches for the 2013 report (though this is a useful long-term NCA goal).

One idea is to use existing case studies that can be compared qualitatively as different templates for local or regional users with four or five different approaches (i.e., good examples of existing scenario exercises) to give people options to pick from and expand upon the range of what they have seen and experienced in the past.

- There are many ways to improve adoption of information developed in assessments through better communication and engagement. One way is to have participatory dialogues and emphasize transparency in the process. To the degree that we can, we should use carefully-prepared visualization techniques to improve clarity of information, engagement, and understanding (Pond et al., 2010).[4] Visualization helps people really grasp scenario concepts.

- Working on solutions is empowering. People often prefer focusing on what they can do, rather than dwelling on the negative impacts of climate change. One of the reactions that stakeholders tend toward is advocacy or writing to politicians. The NCA will have to consider how to balance the issue of empowering people who want to do something without supporting an advocacy program or platform. How might the NCA approach this issue? A common objective is to reduce risk, which is the general approach that the NCA might consider taking.

- The WaterSIM project enables people to toggle through alternative variables or scenarios to see the implications of changing these variables. One can run extreme events based on past experience or even demonstrate water running out before 2023 using specific assumptions. On the demand side, users can show actions that reduce water use outside of the house, how conservation of exterior water changes the water balance more than indoor conservation, etc. Thus, visualization tools can help make information actionable.

Regarding the quantification of uncertainty, the panelists were asked (1) do you consider the handling of it to be inadequate, (2) what would you consider as further development in the context of how uncertainty is represented, and (3) would having probabilistic information help particular studies?

_____

[4] http://www.calp.forestry.ubc.ca/wp-content/uploads/2010/02/CALP-Visioning-Guidance-Manual-Version-1.1.pdf

Responses to these questions highlighted that stake-holders understood how to deal with uncertainty in New York City. The technical committee showed them the IPCC emissions trajectories and the current emissions trajectory. They used three emissions scenarios, giving equal probabilities to A1, B2, and A1B, illustrating that the current trajectory is now above the high end of previous projections.

One option moving forward is developing a guide-book for building scenarios in regional or sectoral assessment meetings in the next nine months. This conversation with stakeholders needs to be structured in realistic and practical ways. Workshop participants suggested providing some illustrations of good approaches, as well as building an inventory of existing examples.

The difficulty of managing the interface between science and policy in using climate model outputs and downscaling was mentioned several times. Proper framing helps decision makers understand unexpected results. Helping them ask the right questions is a worthwhile endeavor, rather than taking their perceived needs for information at face value. The translation function by knowledgeable facilitators adept at participatory processes is a very proactive way to prepare decision makers for a wide range of possible futures but also help them understand that not all futures are equally probable.

3.7 Update: U.S. Global Change Research Program and the NCA

3.7.1 Feedbacks between Assessments and Research

Tim Killeen, U.S. Global Change Research Program
Because it is important to put the Assessment in the broader context of the U.S. Global Change Research Program (USGCRP), Tim Killeen was asked to provide an update on the USGCRP strategic planning process, which he leads. The program is being restructured and a new vision and mission statement have been developed:

• Vision for USGCRP: "A nation, globally engaged and guided by science, meeting the challenges of climate and global change."

• Mission: "To build a knowledge base that informs human responses to climate and global change through coordinated and integrated federal programs of research, education, communication, and decision support."

The new approach includes a strong focus on decision-support activities, based on the best science, including uncertainties and confidence levels and appropriate scales and parameters. It also includes "end-to-end" analysis that provides for feedback from societal users of scientific information, allowing for adjustments to the research agenda to better meet their needs. One end is the users; one end is the information base; and in between is decision support. There is a lot of capacity building required, especially in social science.

In Canada there are National Centers for Excellence to create research networks across the country, and similar networks in Europe and Australia. This is another level of capacity building beyond science and technology centers. There is also a need to find places in the government that will fund applied social science research programs. This is a stumbling block. There may be enough social scientists, but not enough focused on this topic; thus there is a need to reward action and activities in this topic. Within the National Science Foundation, Social Behavioral and Economic Sciences focuses predominantly on cognitive science and not on society's relationship with the Earth.

3.8. Ongoing Scenario Activities: How Can They Support the NCA?

3.8.1 Socioeconomic Baseline Data, Narratives, and Scenarios

Tom Wilbanks, Oak Ridge National Laboratory and Jae Edmonds, Pacific Northwest National Laboratory
The integrated assessment modeling (IAM) community is developing new scenarios as inputs to national, regional, and international processes; some of this work is directly relevant to the next NCA, while other work will be relevant to later assessments. To date, much of the IAM research has focused on building scenarios in the context of greenhouse gas emissions mitigation, but their scope is expanding. The IAM community-based scenario processes underway include the development of Representative Concentration Pathways (RCP), which are now completed. Post-RCP scenario activities are focused on a joint product with the impacts, adaptation, and vulnerability community. Other scenario building activities include the Stanford Energy Modeling Forum; the Asia Modeling Exercise; the RoSE Project; the Program on Integrated Assessment Model Development, Diagnostics and Inter-Model Comparisons (PIAMDDI); the European counterpart

to PIAMDDI, EU FP7, Climate Policy Outreach; and RECIPE and ADAM.

There is other relevant work being performed by individual research groups, including regional disaggregated integrated assessment models, and the development of integrated Earth system models explicitly coupled with components of integrated assessment models.

The RCP process was undertaken to deliver emissions scenarios to the climate modeling community. Some key aspects of the RCPs include development from peer-reviewed literature, harmonization to a common base year, and representation of socioeconomic uncertainties as a set. The RCPs do not include socioeconomic information. The RCP database includes information on forcing agents (emissions, aerosols, derived greenhouse gases, land use, and land cover). Extensions include some downscaling of short-lived species, land use, or land cover.

After the climate model ensembles are complete, pattern scaling could be used to create wholly new socioeconomic and climate change scenarios. The new scenario development process includes the development of storylines and the parallel process. This work is in progress and will include shared socioeconomic pathways that are intended to span ranges of adaptive and mitigative capacity and include socioeconomic variables (e.g., population, macroeconomic activity, and a storyline). There will be five socioeconomic pathways per climate signal (2.6, 4.5, 6.0, and 8.5 Watts per square meter). Many of the details of the socioeconomic pathways are still being worked out, such as the links between mitigation and climate change.

Finally, the PIAMDDI is a Department of Energy-sponsored program designed to improve IAMs through research on technology, uncertainty, impacts and adaptation, regional integrated assessment, and the energy-water-land nexus. It also includes research programs on IAM diagnostics and on IAM model intercomparisons.

3.8.2 Environmental Scenarios—Harmonization of Global Land-Use Scenarios for the Period 1500–2100
George Hurtt, University of Maryland
Land-use change is a central component of the Earth system. Previous efforts in understanding land-use changes have focused on emissions, but efforts are needed to combine the net effects of land-use (e.g., biogeochemical, biophysical, and biodiversity). There is also a need for consistency and cross-disciplinary efforts that can look across timescales, models, spatial resolution, and applications. There have been considerable efforts associated with the AR5 that aim to harmonize efforts, such as developing consensus land-use history reconstructions, minimizing differences between the end of historical reconstructions and the beginning of future projections, and preserving as much information as possible from integrated assessment models regarding the future.

Lessons for the NCA are that modeling the effects of land use in the Earth system need to treat land-use patterns, transitions, management practices, vegetation, biogeochemistry, and biophysics (amongst other considerations) consistently in the past, present, and future. Also, future studies need to harmonize these patterns and transitions, while reducing inconsistencies and preparing for the next generation of fully integrated models.

3.8.3 Climate Data and Scenarios
Karl Taylor, Program for Climate Model Diagnosis and Intercomparison
The Coupled Model Intercomparison Project (CMIP) began in 1995 and is in its fifth phase, which began in 2006. CMIP5 is an ambitious variety of realistic and diagnostic experiments. There are over 20 participating groups with about 40 models. Most results will be available by 2012.

The CMIP5 activity includes new experiments that are designed to be more informative about why models differ in their projections (most between 1° and 2.8° latitude) and are more complete (e.g., include carbon cycle). The model output will consist of more variables of interest to those studying impacts; will have more information on cloud dynamics; and will have output for use for dynamical and statistical downscaling. In addition, CMIP5 will have more complete documentation of models and experiments and a new strategy for delivering model output to users.

The three types of simulations are decadal climate predictions, "long-term" projections (century and longer), and atmosphere-only simulations. The long-term projections design focuses on model evaluation, projections, and improved understanding. Decadal prediction experiments are initialized from the observed climate state (note that the decadal predictions are in an exploratory stage) with the goal of better reproducing the actual climate trajectory in the coming decade or two. The atmosphere-only experiments are targeted for computationally demanding (e.g., high-resolution) models.

CMIP5 output fields are much more comprehensive than those from CMIP3 and will include a wide range of domains and temporal sampling. Daily output may be particularly useful for the NCA. The daily atmospheric and surface variables will include maximum, minimum, and mean surface air temperature, precipitation, humidity, surface wind, and pressure near sea level for all simulations and times. There will be 26 two-dimensional and 7 three-dimensional fields for control runs, historical runs, RCP concentration or emission run years, and AMIP years. In addition, 3-hourly output will also be available for the first time.

CMIP5 output can be used for dynamical downscaling; CORDEX is a WCRP-endorsed effort to do this. All model output will be available for educational and research purposes, and about half of all output will be available for unrestricted use via the Earth System Grid Federation. There will be extensive associated documentation for transparency and improved access.

Given the timelines for the 2013 NCA and CMIP5, the NCA will likely be based primarily on CMIP3 results. CMIP3 results are well-studied and have resulted in more than 500 publications. It may be possible to augment with last-minute additions of some results from CMIP5, but it should be noted that there are different forcing scenarios for CMIP3 and CMIP5, making it difficult to translate between them.

For sustained support of the NCA, the CMIP benchmark experiments provide a basis for quantitative measures of model performance and response, multi-model perspectives, output for directly investigating the consequences of future climate change, and output that can be downscaled for investigating consequences of future climate change.

Uncertainty quantification remains a problem. The value of all climate projections is limited not so much by uncertainty, but by the lack of knowledge about how large the uncertainty is. It is not possible to generally quantify uncertainty in projections. Uncertainty in various major feedbacks needs to be reduced, and the extent to which one should trust the extra detail provided by examining climate

change at ever-smaller scales is not known (this applies to both temporal and spatial scales).

3.8.4 Regional Development of Scenarios: Roles and Capabilities of Boundary Organizations
Philip Mote, Oregon State University
There are currently 11 NOAA-funded Regional Integrated Sciences & Assessments (RISA) teams. As regional teams, they are closely connected with local end uses (variables, scales, and methods), but they are also key links to global-scale science. Innovations that take place at the nexus between local and global spatial scales and short and longer timescales (episodic to decadal) will hopefully lead to research that is seen as more salient, legitimate, and credible by local users. The RISAs are focusing on identifying the timescales of most relevance (e.g., annual and seasonal), sampling global climate model results in a meaningful way (e.g., for regional hydrology and wildfire applications), and variables and seasons of particular interest to the regions (e.g., extreme heat, impacts to permafrost, and seasonal precipitation patterns). RISA teams have also been involved in original research including analyses using regional climate models and developing new statistical downscaling methods.

Model evaluation can focus on special climatological features that are important to a region. Users of regional models often want a continuity of timescales, not just models that project to 2050 and 2100. Also, regional innovations can serve as pilot studies that can then be shared nationally, which is particularly relevant to the NCA.

3.8.5 Post-Presentation Discussion
It would be a mistake to design a rich protocol for building scenarios that would work in every case. There is a need for a participatory process involving both users and scientists to determine what makes sense for regions and sectors, especially for socio-economic and environmental scenarios. This needs to be addressed on a case-by-case basis. One criticism of the SSP process is that it is not as inclusive or "bottom-up" as it could be; it is probably desirable for the NCA to have a more open process. The social science aspects of the SSPs are in progress. There may be qualitative responses by July 2011, and quantitative information may be available later.

Regarding CMIP5, the global climate model output serves a certain segment of the research community, but it is fairly raw. In the future, there will be opportunities for some processing of data to form climatologies that will make it more accessible for a range of users. However, it is difficult to anticipate their needs. If users want socioeconomic data, climate data, and environmental data, then there is a need to go to several different sources. This is not ideal, but perhaps there might be a way to create a directory to the different data sources.

It will be at least a year before CMIP5 data are vetted and ready for use. With CMIP5, land-use changes are being used to drive some of these runs, and land use will be in the historical and future period runs. CMIP3 is well studied and vetted but does not include policy options.

With the capabilities of the NOAA technical support unit at NCDC, the NCA will be building data management capacity to provide archiving and access to data that are developed in real time. At the NCA knowledge management workshop, suggestions were made related to data management, retrieval (metadata), and Web access. Workgroups are being established for all of these topics. There is no intent to create new data systems, rather interoperability and linking to existing data. The intent is to create a directory initially, and in the long term create dynamic integrated support functions. The most important task is to deploy the indicators of change in the next several years, so it will be essential to make sure that the data system can support the indicators. In addition, a clearinghouse and capability mapping function may be included to support the USGCRP adaptation science workgroup's activities. Simple data layers can be put together in the short term, but a more sophisticated plan is needed for the long term. The most important issue is to be able to deploy the data needed for indicators quickly enough so that participants and users can start using them in next several years.

3.9 Communities of Practice Breakout Groups: Next Steps for Producing Scenarios for the NCA

Breakout groups were formed based on communities of practice to discuss options for producing and using different scenario products and resources for both (1) the NCA report due in 2013, and (2) for ongoing assessment activities in regions and sectors. The breakout groups discussed these two topics for the remainder of the second day and for the first

hour of the third day, and then presented their ideas to the larger workshop group.

Group1. Climate Information (e.g., Climate Observations and Modeling Community), Facilitated by Philip Mote, Oregon State University
The overarching feeling within this group was that the NCA should primarily rely on data sets that are already available (e.g., CMIP3 and as available CMIP5), and rely on already available downscaling information and inventories of relevant research (e.g., CCMVal). Participants should be operating under the general ground rule that there is no funding guarantee for these efforts.

The breakout group highlighted several opportunities that the NCA could address and existing or planned efforts that NCA could take advantage of including comprehensive evaluation of climate simulations (e.g., mean and variance and seasonal information); comparison and evaluation of statistical versus dynamical downscaling; extend existing model inventories (e.g., finer resolutions and more variables); perform some CMIP5 versus CMIP3 intercomparisons; apply BCSD (Bias Corrected Statistical Disaggregation) to NARCCAP; provide access to and perform comparisons of existing regional climate model results (including very high-resolution regional climate models); and improve the ease of use of climate models for impacts modelers.

The group also highlighted the need to distinguish, quantify, and discuss sources of uncertainty (e.g., scenarios, within and across models, and natural inter-annual variability). In particular, evaluations and intercomparisons of models are important when characterizing uncertainty. Other areas for addressing uncertainty include promoting consistent treatment of uncertainty across regional and sectoral assessments, using signal-to-noise as a starting point, and exploring different means of characterizing, quantifying, and presenting uncertainty to different audiences. Finally, consistency and a common scope will be critical for producing cross-cutting topic sections of the NCA, for which CMIP3 can be a starting point.

The United Kingdom, Australia, and California have selected a core set of climate simulations to be used in their assessments. What was the process used to select that core set? In California, Dettinger and Cayan mainly consulted with climate scientists. For access and comparisons of regional climate models, we could leverage RISAs or other regional centers to have expert judgment on climate change for each region. NOAA seems to have a lot of interest in doing this—it would add value without doing new runs. Katherine Hayhoe and Don Wuebbles have developed techniques for comparisons of statistical and dynamical downscaling through her dissertation. They do not have funding to do this yet, so a community effort would be needed to do this. Linda Mearns was nominated to lead a group on these near-term opportunities for NCA 2013.

Group 2. Socioeconomic and Environmental Scenarios (e.g., Integrated Assessment and Other Socioeconomic Modelers), Facilitated by Tom Wilbanks, Oak Ridge National Laboratory

The scope of this group was enormous and included socioeconomic scenarios, emissions scenarios, policy scenarios, and environmental scenarios beyond climate modeling. The breakout group included advocates for all of these kinds of scenarios as part of the long-term National Climate Assessment infrastructure. However, their main concerns were not focused on what scenarios to produce, rather on how scenarios are used. It is not yet clear what processes and structures the scenarios will feed into.

One dominant view was that the third NCA should do something different from preceding assessments. It should be the first in important ways; a stepping-stone toward a more coherent national response, not just an application of a familiar formula that has had little impact.

This NCA should be seen as a climate change risk management assessment, not an impact assessment—including adaptation, mitigation, and scientific investment tradeoffs, related to the new RCP scenarios for framing if not for analysis, which paints a different picture of climate change futures than SRES. The NCA should focus on vulnerability scenarios in multiple stresses, multi-hazard contexts, and related to major impact concerns such as water scarcity, food shortages, disaster hot-spots, and ecological thresholds. Again, there should be a focus on "what keeps you awake at night?"

U.S. vulnerabilities and responses should be placed in international contexts (e.g., sulfur emission thresholds in China, international trade, and migration), which require global scenarios as well as national scenarios.

For the long term, the NCA should aspire to become the "gold standard" for climate change response and risk management assessments, by including the following in its approach:

- Socioeconomic narratives and supporting quantitative scenarios (e.g., economic change and associated demands for resource-intensive products and services, demographic change, changes in settlement patterns, especially vulnerable populations, technological change, human resource issues, and changes in institutional and governance structures for decision making and problem solving—related to multiple stressors, adaptation, and mitigation capacities;

- Environmental and ecological scenarios reflecting multiple stresses on land use, biogeochemical dynamics (e.g., nitrogen), vegetation shifts, and disease and pest infestations;

- Scenarios representing inter-sectoral and inter-regional interactions (e.g., land, energy, and water system connections; and demographic and housing growth relative to the preservation of wildlife resources);

- Policy scenarios reflecting both climate policy and development policy, including impacts of policy choices; and

- Effective ways to explore and communicate possible surprises (e.g., in connection with security concerns).

The third NCA scenarios and other activities should be conceived as stepping stones to a more ambitious and more coherent longer-term process. The NCA should focus on advancing capacities to develop appropriate scenarios, by considering links with observational data and identifying needs for new data, and considering how new scenarios should be developed (e.g., links with modeling and with participatory processes, attention to issues of consistency, and public acceptance). There should also be an emphasis on improving capacities to track and analyze uncertainties related to decision criteria and risk management, which is not easy where the science base for long-term projections is weak. Improving linkages between different scales of interest is also needed. The linkages will need to address differences in processes, concerns, and mechanisms for using scenario information. There should also be stronger linkages between processes and phenomena from global to local scales (e.g., markets, institutions, and stresses).

In conclusion, there are several next steps that participants in the breakout group would suggest for consideration by the federal advisory committee:

- By spring of 2011, conduct the inventory of what is already available and identify things to incorporate in the NCA;

- By spring of 2011, identify a manageable number of critical issues and stressor interactions for accelerating impact and response case-study research in order to inform the 2013 Assessment, along with ways to get the job done by early 2012;

- Clarify what we mean by vulnerability scenarios and developing them for use in the 2013 Assessment and report (e.g., what are they and how do we produce them?); and

- Get started mobilizing now in order to meet the longer-term needs for a rich family of scenarios of possible futures for systems, besides the global climate, as essential components of multiple causation and stress assessments.

Finally, there is a need to underline the importance of characterizing uncertainty in the 2013 process. We can use published techniques on propagation of uncertainty, comparing model output to observations, etc. The authors of the 2009 NCA report wanted to underscore the observations of climate change where there was higher certainty, such as the increase in extreme precipitation events. If the NCA is framed in terms of vulnerability, it would be good to focus on observations of vulnerability, to root the conceptualization in terms of what people have seen or experienced. We will have an increasing focus on observations as the NCA process continues, and we will have to think about how scenarios fit in this context. The NCA should get the message out to the agencies that we are doing an inventory of scenarios.

Group 3. Organizations to Bridge between Science and Users (e.g., Those Working with Stakeholders), Facilitated by Holly Hartmann, University of Arizona

One of the key questions within this group was, "How can the NCA manage the diversity of stakeholders, which is essentially the population of the U.S.?" If the NCA develops a typology of users, it might be easier to target products to those users. Within a sector, there can be as much diversity as across sectors (e.g., levels of resources and technical sophistication). Thus, there is a need for flexibility in delivery in order to reach a wide diversity of communities and types of stakeholders. A remaining question is whether we need more flexibility than just organizing around sectors and regions.

It will be challenging for the NCA to provide top-down guidance for maintaining comparability, while also allowing flexibility for regional and sectoral groups. Many in the group did not envision the NCA providing fully integrated scenarios for individual scenario planning. Rather, the role should be to provide a scenario framework for intermediaries to use in a manner that is transparent and straightforward (with respect to assumptions, creation steps, data, benchmarks, and type of scenarios) and that offers examples and guidance.

It will be important for the process to be engaging. In the short term, a consultative role is perhaps most appropriate, while in the long term, a collaborative and/or co-development role will be most appropriate. To make the engagement process most effective the NCA should focus on the following:

- Synthesizing and operationalizing lessons;

- Maintaining realistic expectations, and keeping the best interests of stakeholders in mind;

- Using existing networks that are credible and can provide leadership (e.g., RISAs, state climatologists, extension agents; gatekeeper organizations, professional societies, business and trade organizations, and business schools);

- Drawing from and imprimatur of leadership and charisma and seeking public recognition to elevate the profile of the Assessment;

- Strategically targeting currently less receptive groups to build inroads for future work;

- Capacity building, training, workforce development, and supporting post-engagement (e.g., tools, Web manuals or technology, expert directory, and communities of practice);

- Avoiding overpromising, failing to meet commitments, any hints of regulation (i.e., leading with non-regulatory agencies), and lack of integration; and

- Visualization: designing products that are attractive, engaging, informative, and, where appropriate, addressing emotive issues.

Also in the short term, it will be helpful for the NCA to highlight early adopters, leadership case studies (e.g., how they approached top-down versus flexibility and how they related scenarios to local policies), and recognize sub-regional and cross-regional inter-relationships and sectoral diversity.

The NCA could produce a spectrum of outputs; from quantitative to qualitative, including

- Guidance on format and content of scenario studies (e.g., a cookbook, templates, good practices, lessons learned, and examples to mimic);

- Selection criteria for matching scenarios with interests and qualities that foster utility;

- Evaluation of information systems and tools;

- An electronic atlas of basic geospatial data to fill gaps (e.g., *SimCity* and *MyTown*);

- Integration through mash-ups, Web services, and Web 3.0;

- Monitoring: real-time, distributive, citizen science; and

- Case studies that can be quickly highlighted to help users make quick decisions (e.g., those that take two to three months to write and deploy results).

Specifically in terms of data or information systems that need to be developed, again there are short- and long-term needs. In the short term, emphasis should be placed on identifying, explaining, assessing, rewarding success, and connecting. In the long term, emphasis should be placed on extending, integrating, creating, and supporting. Transparency should remain an imperative focus throughout the data and information gathering and synthesizing process, and the NCA should strive to build a more sustainable information and data system.

Finally, there are timing issues associated with scenario development and potential user needs. Generally, bridging organizations work with what they can get when their stakeholders are ready to engage. There is a critical need to coordinate the coincident release of different scenarios. For

example, locally generated scenarios (e.g., WUCA/PUMA) and national climate scenarios (e.g., NOAA) should be timed together when possible. There is also a need for continuous updating and outreach (e.g., on intermediate timelines, progress, and products). It is also important to note that some work can happen irrespective of NCA commitments (e.g., rapid appraisal of successful efforts, pilots, and case studies), and engagement with professional organizations and establishing liaisons will be important.

3.10 Panel Discussion: Synthesis of Possible Uses, Products, and Options

In this final session of the workshop, speakers offered observations about needs and uses of scenarios for the NCA report to be prepared by 2013 and for the Assessment process in the longer term, highlighting promising options for preparing needed scenario products and indicating research needs. Summaries of remarks by four panelists are provided below.

Kathy Jacobs, White House Office of Science and Technology Policy
The NCA is all about coordination, integration, and serving as the connective tissue between science and decision making. However, there are many things that people want this assessment to be, but it cannot be everything for everyone. It is important to draw from the Adaptation Science Workshop and other existing efforts, as marginal increases in effort can yield great contributions. Regional evaluations of current and future climate model runs, based on expert judgment for regional futures, plus identifying what the regional trends are from stakeholders could represent one of these marginal increases in effort. It will also be important to identify current good-practices and areas that define the state of the science.

We clearly have an issue with communicating uncertainty. We need to own the success of the NCA together and build it together. There will ultimately be funding for these efforts in the future, even if not in the immediate timeframe, but it will require a considerable amount of devotion and patience from the research community in the interim period.

The workshop reports are being written not just as summaries, but as communication documents, which will go to the federal advisory committee. Ending on an encouraging note, Chris West from UKCIP recently said that the approach that the third NCA has taken to doing methodologies and pre-workshops strategies is quite innovative. So, if we get this right, we may set the "gold standard" for conducting National Climate Assessments.

John Hall, Department of Defense

In the past, some of the mission-based (versus science provider) agencies of the federal government, including the Department of Defense, may not necessarily have considered NCA activities as directly applicable to their mission. However, the speaker suggested that they are now beginning to view the NCA and what it can provide in a different light because of the more explicit focus on providing actionable information. In applying this new focus, however, the federal advisory committee will need to come to grips with deciding what is going to be an explicit part of the 2013 Assessment versus what will be part of a visionary approach for the long-term Assessment process. In regards to the 2013 Assessment itself, it would be helpful to establish now an integrated conceptual model of scenarios and models and how they relate. We need to also consider appropriate system boundaries and to what degree scenarios need to be consistent within and across the boundaries of interest to the various users. For example, because of the multitude of military installations spread across the U.S. and its territories, it is important for the DoD to have appropriate regional consistency in the climate scenarios applied by such installations. Finally, it is important to consider the pace of climate-related changes and the ways in which scenarios can inform actions we take now versus those that can wait for future decisions and information.

Jerry Melillo, Marine Biological Laboratory

The NCA must be aware of its dual mission—meeting 2013 and longer-term goals. The 2013 report is the "art of the possible", but the longer term is about getting on the road to the "gold standard." To accomplish this, we need to make 2013 part of the transition (i.e., do not do anything for 2013 that leads to a dead end). An overall "vulnerability framing" should work as long as it does not preclude analysis of opportunities and solutions.

There are several challenges that the NCA faces (e.g., getting it "just right", or having enough guidance but also enough flexibility, and the need for constant communication and interaction with participants if we need to adapt and implement a "Plan B"). A general path forward might include

- Taking advantage of ongoing activities;

- Making a clear connection to the international context;

- Identifying case studies that consider both opportunity and vulnerability;

- Considering crosscuts, examine the obvious list carefully;

- Pursuing near-term action items, such as (1) inventories of scenario products and activities using scenarios—ask agencies and key players in the research community; (2) commissioning a few targeted efforts; and (3) scenarios for NCA 2013 should provide inputs into development of the long-term process, including revision and publication of the scenarios white paper (Chapter 2);

- Including observations in scenario formulation;

- Thinking toward the long-term goals by expanding a network interacting and using scenarios in participatory processes, especially in the business and military sectors; and

- Building human capacity and enhancing capacity across sectors and regions by setting up mechanisms to create scenarios and discussing their implications, including building visualization capacity.

Rich Richels, Electric Power Research Institute

There are several points that the NCA 2013 should consider:

- It is critical at the outset to communicate the purpose of scenarios, as there are currently many common misperceptions, particularly that scenarios are predictions. To the contrary, they represent a series of "what if" questions necessary for risk management. Risk management must be based on estimates of the stakes, the odds, and societal attitudes toward risk; this cannot be done without scenarios. Even so, scenarios will nevertheless be controversial. This comes with the territory.

- The risk management process must adapt to new information by modifying scenarios when appropriate. In *America's Climate Choices* (NRC, 2011), this was characterized as iterative risk management.

- In choosing scenarios, we do not want to focus solely on most likely scenarios or the scenario with the greatest damages. Combining information on the stakes and the odds is required (the probability of the scenarios multiplied by the projected damage determines the choices). How far we go down the tails of the distribution is a very important design decision.

- There are many kinds of decisions that need to be made with regard to climate change. The workshop has focused primarily on adaptation decisions, a very important category to all levels of government as well as the private sector. But there are a variety of other decisions that need to be informed, including decisions to slow the rate of climate change; decisions on research and development to develop new technologies on both the supply and demand sides of the energy system and to enhance adaptive capacity; and decisions to fund science so we can make better informed decisions in the future. Scenarios are needed to evaluate and prioritize investments in all of these areas. They are also needed for prioritizing investments across areas.

- The speaker suggested that it would be a major oversight not to examine policy scenarios (particularly with respect to mitigation) in addition to non-policy scenarios. We should be explicit about the costs and benefits of policy measures and the accompanying uncertainties. Actions to curb the rate of climate change may not be a free lunch, but they may turn out to be a lunch well worth buying. People will want to know whether an investment is worth making.

- A caveat exists: the parallel process is a major step forward, but complicates the use of analysis to look at the tradeoffs at the margin between mitigation costs and damages avoided.

- Finally, we need to have a Plan B. If considerably less money is available for the Assessment than currently anticipated, how will the community proceed in producing the required National Climate Assessment?

3.11 Adjourning Comments from the Workshop Co-Chairs

Richard Moss, Pacific Northwest National Laboratory and Linda Mearns, National Center for Atmospheric Research

Key outcomes from the workshop were reported to the group. The complexity of relationships among different types of scenarios and models was noted along with the need to adopt clear and consistent terminology in referring to scenarios. Scenarios are a useful way to support and coordinate multi-sectoral and regional analyses in a fashion that nests them in the context of a range of possible futures in the Earth system, socioeconomic, and environmental conditions. There will need to be a focus on a simple formula for scenarios for the 2013 report, with a more ambitious longer-term goal of developing additional methods and processes to support participatory processes. In addition, there is a strong interest in ensuring that scenario activities of the NCA be use-driven and useful in real-world contexts as well as responsive to communication needs. This implies a broad toolbox of alternative approaches based on an inventory and analysis of what already exists. There are many scenario-based activities currently in place, but they are extremely diverse and oriented toward a wide range of users—some of whom are end users of information and others who are intermediate users or translators of information between scientists and decision makers. While the workshop identified a number of promising options, it did not resolve many detailed issues that will need to be determined by the federal advisory committee responsible for the Assessment.

REFERENCES

Arnstein, S. R. 1969. A ladder of citizen participation. *Journal of the American Planning Association* **35**:216-224.

Bradfield, R., G. Wright, G. Burt, G. Cairns, and K. van der Heijden. 2006. The origins and evolution of scenario techniques in long range business planning. *Futures* **37**:795-812.

Bryan, B. A., C. M. Raymond, N. D. Crossman, and D. King. 2011. Comparing spatially explicit ecological and social values for natural areas to identify effective conservation strategies. *Conservation Biology* **25**:172-181.

Burch, S., R. J. Sheppard, A. Shaw, and D. Flanders. 2010. Planning for climate change in a flood-prone community: Municipal barriers to policy action and the use of visualizations as decision-support tools. *Journal of Flood Risk Management* **3**:126-139.

Carter, T. R., E. L. La Rovere, R. N. Jones, R. Leemans, L. O. Mearns, N. Nakicenovic, A. B. Pittock, S. M. Semenov, and J. Skea. 2001. Developing and applying scenarios. Pages 170-190 *in* J. J. McCarthy, O. F. Canziani, N. A. Leary, D. J. Dokken, and K. S. White, editors. Climate Change 2001. Cambridge University Press, Cambridge, U.K.

CCSP. 2003. Strategic Plan for the Climate Change Science Program. Climate Change Science Program, Washington, D.C.

Desai, S. and M. Hulme. 2004. Does climate adaptation policy need probabilities? *Climate Policy* **4**:107-128.

EEA. 2009. Looking Back on Looking Forward: A Review of Evaluative Scenario Literature. European Environment Agency, Copenhagen, Denmark.

Fischer, F. 1993. Citizen participation and the democratization of policy expertise: From theoretical inquiry to practical cases. *Policy Sciences* **26**:165-187.

Fisher, B. S., N. Nakicenovic, K. Alfsen, J. C. Moriot, F. de la Chesnaye, J.-C. Hourcade, K. Jiang, M. Kainuma, E. La Rovere, A. Matysek, A. Rana, K. Riahi, R. Richels, S. Rose, D. Van Vuuren, and R. Warren. 2007. Issues related to mitigation in the long-term context. Pages 169-250 *in* B. Metz, O. R. Davidson, P. R. Bosch, R. Dave, and L. A. Meyer, editors. Climate Change 2007: Mitigation of Climate Change. Cambridge University Press, Cambridge, U.K.

Gonzalez, P., R. P. Neilson, J. M. Lenihan, and R. J. Drapek. 2010. Global patterns in the vulnerability of ecosystems to vegetation shifts due to climate change. *Global Ecology and Biogeography* **19**:755-768.

Grubler, A. and N. Nakicenovic. 2001. Identifying dangers in an uncertain climate. *Nature* **412**:15.

Hall, J. 2007. Probabilistic climate scenarios may misrepresent uncertainty and lead to bad adaptation decisions. *Hydrological Processes* **21**:1127-1129.

Hawkins, E. and R. Sutton. 2010a. The potential to narrow uncertainty in projections of regional precipitation change. *Climate Dynamics* **37**:407-418.

Hawkins, E. and R. Sutton. 2010b. The potential to narrow uncertainty in regional climate predictions. Bulletin of the American Meteorological Society **90**:1095.

Karl, T. R., J. M. Melillo, and T. C. Peterson, editors. 2009. Global Climate Change Impacts in the United States. Cambridge University Press, Cambridge, U.K.

Katz, R. W. 2002. Techniques for estimating uncertainty in climate change scenarios and impact studies. *Climate Research* **20**:167-185.

Knutti, R., F. Joos, S. A. Muller, G.-K. Plattner, and T. F. Stocker. 2005. Probabilistic climate change projections for CO_2 stabilization profiles. *Geophysical Research Letters* **32**:L20707.

Kriegler, E., B. C. O'Neill, S. Hallegatte, K. T., R. H. Moss, R. Lempert, and T. J. Wilbanks. 2010. Socio-economic Scenario Development for Climate Change Analysis. Centre International de Recherche sur l'Environnement et le Développement, Nogent-sur-Marne Cedex, France

Loibl, W. and A. Walz. 2010. Generic regional development strategies from local stakeholders' scenarios - the Montafon experience. *Ecology and Society* **15**(3).

MacCracken, M. 2000. Status report and some initial thoughts on lessons learned from the first phase of the U.S. National Assessment on the potential consequences of climate variability and change. Washington, D.C.

MacCracken, M., E. Barron, D. Easterling, B. Felzer, and T. Karl. 2001. Scenarios for climate variability and change. Pages 13-72 *in* NAST, editor. Climate Change Impacts on the United States. Cambridge University Press, Cambridge, U.K.

Mearns, L. O., M. Hulme, T. R. Carter, R. Leemans, M. Lal, and P. Whetton. 2001. Climate scenario development. Pages 739-768 *in* J. T. Houghton, Y. Ding, D. J. Griggs, M. Noguer, P. J. van der Linden, X. Dai, K. Maskell, and C. A. Johnson, editors. Climate Change 2001: The Scientific Basis. Cambridge University Press, Cambridge, U.K.

Melillo, J., A. Janetos, D. Schimel, and T. Kittel. 2001. Vegetation and biogeochemical scenarios.in NAST, editor. Climate Change Impacts on the United States: The Potential Consequences of Climate Variability and Change. Cambridge University Press, Cambridge, U.K.

Morgan, M. G., R. Cantor, W. C. Clark, A. Fisher, H. D. Jacoby, A. C. Janetos, A. P. Kinzig, J. Melillo, R. B. Street, and T. J. Wilbanks. 2005. Learning from the U.S. National Assessment of climate change impacts. *Environmental Science and Technology* **39**:9023-9032.

Moss, R. and J. Marengo. 2007. Letter to PCMDI regarding TGICA-PCMDI cooperation on facilitating access to a subset of the IPCC model data archive, Annex 1: List of AR4 variables. TGICA10/Doc. 6. http://www.ipcc.ch/pdf/activity/doc-11-tgica.pdf.

Moss, R., M. Babiker, S. Brinkman, E. Calvo, T. Carter, J. Edmonds, I. Elgizouli, S. Emori, L. Erda, K. Hibbard, R. Jones, M. Kainuma, J. Kelleher, J. F. Lamarque, M. Manning, B. Matthews, K. Riahi, S. Rose, P. Runci, R. Stouffer, D. van Vuuren, J. Weyant, T. Wilbanks, J. P. van Ypersele, and M. Zurek. 2008. Towards New Scenarios for Analysis of Emissions, Climate Change, Impacts, and Response Strategies. Intergovernmental Panel on Climate Change, Geneva, 132 pp.

Moss, R. H., J. A. Edmonds, K. Hibbard, M. Manning, S. K. Rose, D. Van Vuuren, T. R. Carter, S. Emori, M. Kainuma, T. Kram, G. A. Meehl, J. F. B. Mitchell, N. Nakicenovic, K. Riahi, S. J. Smith, R. J. Stouffer, A. M. Thomson, J. P. Weyant, and T. J. Wilbanks. 2010. The next generation of scenarios for climate change research and assessment. *Nature* **463**:747-756.

Nakicenovic, N. and R. Swart, editors. 2000. Special Report on Emissions Scenarios. Cambridge University Press, Cambridge, U.K.

Nakicenovic, N., J. Alcamo, G. Davis, B. de Vries, J. Fenhann, S. Gaffin, K. Gregory, A. Grubler, T. Y. Jung, T. Kram, E. L. La Rovere, L. Michaelis, S. Mori, T. Morita, W. Pepper, H. Pitcher, L. Price, K. Riahi, A. Roehrl, H.-H. Robner, A. Sankovski, M. Schlesinger, P. Shukla, S. Smith, R. Swart, S. van Rooijen, N. Victor, and Z. Dadi. 2000. Scenario driving forces. Pages 103-166 in N. Nakicenovic and R. Swart, editors. Special Report on Emissions Scenarios. Cambridge University Press, Cambridge, U.K.

NAST, editor. 2001. Climate Change Impacts on the United States: The Potential Consequences of Climate Variability and Change. Cambridge University Press, Cambridge, U.K.

Nicholson-Cole, S. 2005. Representing climate change futures: A critique on the use of images for visual communication. *Computers, Environment and Urban Systems* **28**:255-273.

NRC. 2007. Analysis of Global Change Assessments: Lessons Learned. Committee on Analysis of Global Change Assessments, National Research Council, National Academies Press, Washington, D.C.

NRC. 2009. Informing Decisions in a Changing Climate. Panel on Strategies and Methods for Climate-Related Decision Support, National Research Council, The National Academies Press, Washington, D.C.

NRC. 2010. Describing Socioeconomic Futures for Climate Change Research and Assessment: Report of a Workshop. Panel on Socio-Economic Scenarios for climate change Research and Assessment, National Research Council, The National Academies Press, Washington, D.C.

NRC. 2011. America's Climate Choices. Committee on America's Climate Choices, National Research Council, National Academies Press, Washington, D.C.

Parson, E. A., V. Burkett, K. Fisher-Vanden, D. Keith, L. O. Mearns, H. Pitcher, C. Rosenzweig, and M. Webster. 2007. Global Change Scenarios: Their Development and Use. Climate Change Science Program, Washington, D.C.

Parson, E. A., M. G. Morgan, A. Janetos, L. Joyce, B. Miller, R. Richels, and T. J. Wilbanks. 2001. The socioeconomic context for climate impact assessment. Pages 93-107 *in* NAST, editor. Climate Change Impacts on the United States: The Potential Consequences of Climate Variability and Change. Cambridge University Press, Cambridge, U.K.

PCMDI. 2011. CMIP5 - Modeling Info - Producing Model Output. CMIP5 Coupled Model Intercomparison Project, World Climate Research Programme. http://www-pcmdi.llnl.gov/

Pittock, A. B., R. N. Jones, and C. D. Mitchell. 2001. Probabilities will help us plan for climate change. *Nature* **413**:249.

Pond, E., O. Schroth, S. R. J. Sheppard, S. Muir-Owen, I. Liepa, C. Campbell, J. Salter, K. Tatebe, and D. Flanders. 2010. Local Climate Change Visioning and Landscape Visualizations: Guidance Manual. Collaborative for Advanced Landscape Planning, University of British Columbia, Vancouver, Canada.

Robinson, J. 2008. Being undisciplined: Transgressions and intersections in academia and beyond. *Futures* **40**:70-86.

Ruosteenoja, K., T. R. Carter, K. Jylha, and H. Tuomenvirta. 2003. Future Climate in World Regions: An Intercomparison of Model-Based Projections for the New IPCC Emissions Scenarios. Finnish Environment Institute, Finnish Meteorological Institute, Helsinki, Finland.

Salter, J. D., C. Campbell, M. Journeay, and S. R. J. Sheppard. 2009. The digital workshop: Exploring the use of interactive immersive visualisation tools in participatory planning. *Journal of Environmental Management* **90**:2090-2101.

Schneider, S. H. 2001. What is "dangerous" climate change? *Nature* **411**:17-19.

Sharp, D. 2010. An inventory of approaches to climate modeling and downscaling. Presentation at PUMA Workshop, San Francisco, California, December, 2010.

Shaw, A., S. R. J. Sheppard, S. Burch, D. Flanders, and A. Wjek. 2009. Making local futures tangible: Synthesizing, downscaling, and visualizing climate change scenarios for participatory capacity building. *Global Environmental Change* **19**:447-463.

Sheppard, S. R. J. 2005. Landscape visualization and climate change: The potential for influencing perceptions and behaviour. *Environmental Science and Policy* **8**:637-654.

Sheppard, S. R. J., A. Shaw, S. Burch, D. Flanders, A. Wiek, J. Carmichael, J. Robinson, and S. Cohen. 2011. Future visioning of local climate change: A framework for community engagement and planning with scenarios and visualization. *Futures* **43**:400-412.

van Notten, P. W. F., J. Rotmans, A. van Asselt, and D. A. Rothman. 2003. An updated scenario typology. *Futures* **35**:423-443.

van Vuuren, D. P. 2010. Developing new scenarios as a common thread for future climate research. Supporting material for the IPCC Workshop on Socioeconomic Scenarios, Berlin, November 2010. http://www.ipcc-wg3.de/meetings/expert-meetings-and-workshops/WoSES.

van Vuuren, D. P., J. Feddema, J.-F. Lamarque, K. Riahi, S. Rose, S. Smith, and K. Hibbard. 2008. Work plan for data exchange between the integrated assessment and climate modeling community in support of Phase-0 of scenario analysis for climate change assessment (Representative Community Pathways), http://cmip-pcmdi.llnl.gov/cmip5/docs/RCP_handshake.pdf.

Vervoort, J. M., K. Kok, R. Van Lammeren, and T. Veldkamp. 2010. Stepping into futures: Exploring the potential of interactive media for participatory scenarios on social-ecological systems. *Futures* **42**:604-616.

Weyant, J., O. R. Davidson, H. Dowlatabadi, J. A. Edmonds, M. Grubb, E. A. Parson, R. Richels, J. Rotmans, P. R. Shukla, R. S. J. Tol, W. Cline, and S. Fankhauser. 1996. Integrated assessment of climate change: An overview and comparison of approaches and results. Pages 367-396 *in* J. P. Bruce, H. Lee, and E. F. Haites, editors. Climate Change 1995: Economic and Social Dimensions of Climate Change. Cambridge University Press, Cambridge, U.K.

Wiek, A., L. Withycombe-Keeler, and R. Kutter. 2010. Participatory approaches for constructing and using the new generation of climate change scenarios - A methodological review. School of Sustainability, Arizona State University, Tempe, Arizona.

Wiek, A., C. Binder, and R. W. Scholz. 2006. Functions of scenarios in transition processes. *Futures* **38**:740-766.

Winn, W. 1997. The Impact of Three-Dimensional Immersive Virtual Environments on Modern Pedagogy. University of Washington, Seattle, Washington.

Zurek, M. and T. Henrichs. 2007. Linking scenarios across geographical scales in international environmental assessments. *Technological Forecasting and Social Change* **74**:1281-1295.

APPENDIX A. WORKSHOP ORGANIZING COMMITTEES

A.1 Research Community Planning Committee

Linda Mearns (Co-Chair), Senior Scientist, National Center for Atmospheric Research; Director, Institute for the Study of Society and Environment

Richard Moss (Co-Chair), Senior Research Scientist, Joint Global Change Research Institute of the Pacific Northwest National Laboratory; Visiting Senior Research Scientist, Earth Systems Science Interdisciplinary Center of the University of Maryland

Molly Cross, Climate Change Adaptation Coordinator, North America Program of the Wildlife Conservation Society

Nathan Engle, Post-Doctoral Researcher, Pacific Northwest National Laboratory

Holly Hartmann, Director, Arid Lands Information Center, University of Arizona

Kathy Hibbard, Manager, Pacific Northwest National Laboratory

Robert Lempert, Director, Frederick S. Pardee Center for Longer Range Global Policy and the Future Human Condition; Professor, Pardee RAND Graduate School

Philip Mote, Director, Oregon Climate Change Research Institute, Oregon State University; Professor, College of Oceanic and Atmospheric Sciences, Oregon State University

Ted Parson, Joseph L. Sax Collegiate Professor of Law and Professor of Natural Resources and Environment, University of Michigan

Richard Richels, Senior Technical Executive for Global Climate Change Research, Electric Power Research Institute

John Robinson, Executive Director of the University of British Columbia Sustainability Initiative and Professor in the Institute for Resources, Environment and Sustainability and Department of Geography, University of British Columbia

Cynthia Rosenzweig, Senior Research Scientist, NASA Goddard Institute for Space Studies and Columbia Earth Institute; Head of Climate Impacts Group; Professor, Barnard College

Joel Smith, Principal, Stratus Consulting

Tom Wilbanks, Group Leader and Corporate Research Fellow, Global Change and Developing Countries Programs, Environmental Sciences Division, Oak Ridge National Laboratory

A.2 Federal Coordinating Committee

Bob Vallario, Program Manager, Integrated Assessment Research Program, Climate Change Research Division, Office of Science, U.S. Department of Energy (Lead Agency for the Workshop)

John Balbus, Senior Advisor for Public Health, National Institute of Environmental Health Sciences, National Institutes of Health

Anne Grambsch, Acting Staff Director, Global Change Research Program, National Center for Environmental Assessment, U.S. Environmental Protection Agency

John Hall, Program Manager, Resource Conservation and Climate Change, Strategic Environmental Research and Development / Environmental Security Technology Certification Program, U.S. Department of Defense

Kathy Jacobs, Director, National Climate Assessment; Assistant Director for Climate Assessment and Adaptation, White House Office of Science and Technology Policy, Executive Office of the President

Tom Karl, Director, National Climatic Data Center, National Oceanic and Atmospheric Administration; Chair, Subcommittee on Global Change Research

Ken Kunkel, Senior Scientist and Science Lead for Assessments, Cooperative Institute for Climate and Satellites, National Climatic Data Center, National Oceanic and Atmospheric Administration

Linda Langner, National Program Leader, Resources Planning Act Assessment, USDA Forest Service

Sheila O'Brien, Coordinator, National Climate Assessment, U.S. Global Change Research Program

Bob O'Connor, Program Director, Social, Behavioral and Economic Sciences, National Science Foundation

Anne Waple, Program Manager, Assessment Services, National Climatic Data Center, National Oceanic and Atmospheric Administration

APPENDIX B. WORKSHOP AGENDA

Scenarios for National Climate Assessment (NCA):
Supporting the 2013 NCA Report and an Ongoing Assessment Process

Monday, 6 December 2010

8:30 a.m.
  Welcome and Introduction, Introductions, and Objectives of the Workshop:
    Richard Moss (Pacific Northwest National
    Lab/Joint Global Change Research Institute) and Linda Mearns (National Center for
    Atmospheric Research)

  Workshop Charge and Coordination Process:
    Robert Vallario (Department of Energy)

8:50 a.m.
  NCA Objectives, Structure, and Context
    John Hall (Department of Defense), Chair

  Overview of National Assessment: Needs and Objectives, Process and Organization
    Kathy Jacobs (Office of Science and Technology Policy)

  Scenarios in Global Assessments -- IPCC
    Chris Field (IPCC WG II), remote presentation

  Discussion

9:45 a.m.
  Break

10:00 a.m.
  Types and Uses of Scenarios
    Chair, Richard Moss (PNNL/JGCRI)

  Overview of Scenarios in Climate Research and Assessment
    Edward Parson (University of Michigan)

  Climate Scenarios and Information
    Linda Mearns (NCAR)

  Anticipation of Decadal Prediction Experiments for use in Scenarios
    Lisa Goddard (International Research Institute for Climate and Society (IRI))

  Scenarios in Prior National Assessments
    Thomas Wilbanks (Oak Ridge National Laboratory)

  Nested Scenarios: Approaches for Linking Different Scales of Analysis
    Brian O'Neill (NCAR)

  Discussion

12:30 p.m.
  Lunch

1:30 p.m.
Regions, Sectors, and Crosscutting Issues—Assessment Priorities and Scenario Needs
Chair, Cynthia Rosenzweig (NASA Goddard Institute for Space Studies)

Report from Sectors/Regions Workshop
Kate White (U.S. Army Corps of Engineers)

Report on User Needs Identified in the Piloting Utility Modeling Applications Workshop
David Behar (San Francisco Public Utilities Commission)

Application of Climate Change Scenarios to Natural Resource Management
Patrick Gonzalez (National Park Service)

Interactions of Sectors and Regions: Emerging Use of Scenarios in Analysis of Climate-Security Concerns
Blake McBride (Department of Defense)

Discussion

3:00 p.m.
Break

3:15 p.m.
Breakout Groups I: Possible Uses of and Needs for Scenarios in the NCA

Facilitators: David Behar (San Francisco Public Utilities Commission), Anne Waple (NOAA), Bill Easterling (Penn State University)

Three parallel breakout groups will discuss needs for and uses of scenarios in the NCA. Groups need to consider:

1. Different desired attributes of climate information (contextual information, observations, and
   projections) for NCA 2013 report and ongoing sustainable process
2. Needs for socioeconomic narratives and scenarios (qualitative descriptions of different socioeco
   nomic development pathways) at what scales and focused on what attributes of the future for the NCA 2013 report and ongoing sustainable process?
3. Needs and uses of environmental scenarios (e.g., sea level rise, air quality, water quality/availability, land use) for the NCA 2013 report and ongoing sustainable process
4. Needs for data and information handing—how should scenarios be provided to both intermediate users and stakeholders for the NCA 2013 report and the ongoing sustainable process?

Participants will be provisionally distributed across breakout groups to ensure balanced representation of expertise. If you strongly wish to join another group, please check with the relevant facilitators.

6:00 p.m.
Adjourn

Tuesday, 7 December 2010

8:30 a.m.
Breakout Groups I (cont.): Finalize input on needs and options

9:30 a.m.
Report from Breakout Groups

Break

10:30 a.m.
Participatory Scenario Processes
Chair: Holly Hartmann (University of Arizona)

Scenarios in Participatory Processes: Connecting Stakeholder Concerns to Global Change
Gregg Garfin (University of Arizona)

Use of Scenarios in Metro NYC Assessments and Planning
Cynthia Rosenzweig (NASA Goddard Institute for Space Studies)

Making local futures tangible—synthesizing, downscaling, and visualizing climate change scenarios for participatory capacity building
Stephen Sheppard (University of British Columbia)

Connecting Stakeholders to Global Change Futures: Gaming, Visualization, and Other Methods
James Buizer (Arizona State University)

Discussion

12:00 p.m.
Lunch

1:00 p.m.
U.S. Global Change Research Program and the NCA
Feedbacks between Assessments and Research: NCA and the USGCRP
Tim Killeen (U.S. Global Change Research Program)

1:30 p.m.
Ongoing Scenario Activities: How Can They Support the NCA?
Chair: Bob Chen (CIESIN)

Socioeconomic Baseline Data, Narratives, and Scenarios
Joel Smith (Stratus Consulting)

Environmental Scenarios
George Hurtt (University of Maryland)

Climate Data and Scenarios
Karl Taylor (Program for Climate Model Diagnosis and Intercomparison (PCMDI))

Boundary Organizations and Support for User Interactions
  Phil Mote (Oregon State University)

Discussion

4:00 p.m.
  Communities of Practice Breakout Groups: Next Steps for Producing Scenarios for the NCA

Breakout groups will be formed based on "communities of practice" to discuss options
for producing and using different scenario products and resources for both (1) the NCA
report due in 2013, and (2) for ongoing assessment activities in regions and sectors.

Proposed breakout groups:

1) **Climate information (e.g., climate observations and modeling community) –**
    Facilitator: Phil Mote (Oregon State University)
2) **Socioeconomic and environmental scenarios (e.g., integrated assessment and other
    socioeconomic modelers) –**
    Facilitator: Tom Wilbanks (ORNL)
3) **Organizations to bridge between science and users (e.g., those working with
    stakeholders) –**
    Facilitator: Holly Hartmann

6:00 p.m.
  Adjourn

Wednesday, 8 December 2010

8:30 a.m.
  Breakout Group Reports

9:30 a.m.
  Panel Discussion: Synthesis of Possible Uses, Products, and Options
    Chair: Robert Vallario (DOE)

The speakers will offer observations about needs and uses of scenarios for the NCA report to be prepared
by 2013 and for the assessment process in the longer term. They will highlight promising options for
preparing needed scenario products and indicate research needs. Following these initial presentations,
participants will be asked to comment on a series of issues raised by the discussion moderator, and there
will also be a period for open-ended discussion. This session will draw on all previous sessions.

  Panelists: Kathy Jacobs (OSTP), John Hall (DoD), Jerry Melillo (MBL) Richard Richels (EPRI)

  Discussion

11:30 a.m.

APPENDIX C. PARTICIPANT LIST

John Antle, Oregon State University

Jeffrey Arnold, U.S. Army Corps of Engineers, Institute for Water Resources

Ines Azevedo, Carnegie Mellon University

John Balbus, U.S. Department of Health and Human Services

Dan Barrie, University of Maryland

David Behar, San Francisco Public Utilities Commission

Levi Brekke, U.S. Bureau of Reclamation

Jim Buizer, Arizona State University

Lawrence Buja, National Center for Atmospheric Research

Robert Chen, Center for International Earth Science Information Network, Columbia University

Molly Cross, Wildlife Conservation Society

Bill Easterling, Penn State University

Jae Edmonds, Joint Global Change Research Institute, Pacific Northwest National
Lab/University of Maryland

Nathan Engle, Joint Global Change Research Institute, Pacific Northwest National
Lab/University of Maryland

Gregg Garfin, University of Arizona

Pat Gober, Arizona State University

Lisa Goddard, Columbia University

Bryce Golden Chen, U.S. Global Change Research Program

Patrick Gonzalez, U.S. National Park Service

Anne Grambsch, U.S. Environmental Protection Agency

Lisa Graumlich, University of Washington

John Hall, U.S. Department of Defense

Holly Hartmann, University of Arizona

Kathy Hibbard, Pacific Northwest National Laboratory

George Hurtt, University of Maryland and the Joint Global Change Research Institute

Kathy Jacobs, White House Office of Science and Technology Policy

Tony Janetos, Joint Global Change Research Institute, Pacific Northwest National
Lab/University of Maryland

Milind Kandlikar, University of British Columbia

David Kaufman, U.S. Department of Homeland Security/Federal Emergency
Management Agency

Melissa Kenney, National Oceanic and Atmospheric Administration

Tim Killeen, National Science Foundation

Bob Kopp, U.S. Department of Energy

Linda Langner, U.S. Department of Agriculture, Forest Service

Robert Lempert, RAND Corporation

L. Ruby Leung, Pacific Northwest National Laboratory

Fred Lipschultz, U.S. Global Change Research Program

Mike MacCracken, Climate Institute

Blake McBride, U.S. Department of Defense

Patrick McCarthy, The Nature Conservancy

Chad McNutt, National Integrated Drought Information System

Linda Mearns, National Center for Atmospheric Research

Jerry Melillo, Marine Biological Laboratory

Mark Meo, Southern Climate Impacts Planning Program

Richard Moss, Joint Global Change Research Institute, Pacific Northwest National
Lab/University of Maryland

Philip Mote, Oregon State University

Sheila O'Brien, U.S. Global Change Research Program

Brian O'Neill, National Center for Atmospheric Research

Ted Parson, University of Michigan

Marc Perry, U.S. Census Bureau

John Reilly, Massachusetts Institute of Technology

Richard Richels, Electric Power Research Institute

Steve Rose, Electric Power Research Institute

Cynthia Rosenzweig, National Aeronautics and Space Administration, Goddard Institute for
  Space Studies

Matthias Ruth, University of Maryland

Anji Seth, University of Connecticut

Stephen Sheppard, University of British Columbia

Benjamin Sleeter, U.S. Geological Survey

Joel Smith, Stratus Consulting

Courtney St. John, U.S. Department of Defense

Susan Stewart, U.S. Department of Agriculture, Forest Service

Ron Stouffer, National Oceanic and Atmospheric Administration

Karl Taylor, Lawrence Livermore National Laboratory

June Thormodsgard, U.S. Geological Survey

Bob Vallario, U.S. Department of Energy

Anne Waple, National Oceanic and Atmospheric Administration

Kathleen White, U.S. Army Corps of Engineers

Tom Wilbanks, Oak Ridge National Laboratory

Donald Wuebbles, University of Illinois

Brent Yarnal, Penn State University

David Yates, National Center for Atmospheric Research